LICHENS
OF
BRITAIN & IRELAND

An Introductory Guide

Rebecca Yahr & Frances Stoakley

BLOOMSBURY WILDLIFE
LONDON · OXFORD · NEW YORK · NEW DELHI · SYDNEY

DEDICATED TO ALL THE SMALL ACTS OF COLLABORATION AND CONNECTION THAT MAKE THE WORLD A BETTER PLACE.

BLOOMSBURY WILDLIFE
Bloomsbury Publishing Plc
50 Bedford Square, London, WC1B 3DP, UK
Bloomsbury Publishing Ireland Limited,
29 Earlsfort Terrace, Dublin 2, D02 AY28, Ireland

BLOOMSBURY, BLOOMSBURY WILDLIFE and the Diana logo are trademarks of
Bloomsbury Publishing Plc

First published in the United Kingdom 2025

Copyright © Rebecca Yahr and Frances Stoakley, 2025
Illustrations © Liz Campbell, 2025

Rebecca Yahr and Frances Stoakley have asserted their right under the Copyright, Designs and Patents Act, 1988, to be identified as Authors of this work

For legal purposes the Acknowledgements on pp. 247 and 248 constitute an extension of this copyright page

All rights reserved. No part of this publication may be: i) reproduced or transmitted in any form, electronic or mechanical, including photocopying, recording or by means of any information storage or retrieval system without prior permission in writing from the publishers; or ii) used or reproduced in any way for the training, development or operation of artificial intelligence (AI) technologies, including generative AI technologies. The rights holders expressly reserve this publication from the text and data mining exception as per Article 4(3) of the Digital Single Market Directive (EU) 2019/790

Bloomsbury Publishing Plc does not have any control over, or responsibility for, any third-party websites referred to or in this book. All internet addresses given in this book were correct at the time of going to press. The authors and publisher regret any inconvenience caused if addresses have changed or sites have ceased to exist, but can accept no responsibility for any such changes

The Royal Botanic Garden Edinburgh (RBGE) logo is reproduced with the permission of RBGE

The British Lichen Society typographical font style is used with the permission of the British Lichen Society.
All rights reserved

A catalogue record for this book is available from the British Library

Library of Congress Cataloguing-in-Publication data has been applied for

ISBN: PB: 978-1-3994-0474-7; ePUB: 978-1-3994-0473-0; ePDF: 978-1-3994-0472-3

2 4 6 8 10 9 7 5 3 1

Maps by Martin Brown
Typeset in Brandon Grotesque by D & N Publishing, Wiltshire, UK
Printed and bound in China by RR Donnelley Asia Printing Solutions Ltd, Dongguan, Guangdong

MIX
Paper | Supporting responsible forestry
FSC
www.fsc.org FSC® C144853

To find out more about our authors and books visit www.bloomsbury.com and sign up for our newsletters

For product safety related questions contact productsafety@bloomsbury.com

CONTENTS

Who is this book for?	4
The basics: lichens 101	6
The beautiful intricacy of lichens	18
Lichen identification	24
How to use this field guide	42
The pathways	44
Lichens by habitat	44
A Baker's Dozen of important lichen groups	64
Lichens by form and colour	85
Troubleshooting	234
Resources and next steps	239
Further reading	242
Glossary	244
Photo credits	247
Acknowledgements	248
Index	249

WHO IS THIS BOOK FOR?

We wanted to write this field guide because when we have taught people who are starting out in the wonderful world of lichens, the inevitable question has been: Is there a good beginner's book on the subject? We never felt that we had the right answer, and so we really hope that this fills the gap – offering a first introduction and at the same time paving the way to the many excellent books and resources that already exist but which are not aimed at new learners.

This guide is intended for those who have just begun to look at lichens and want to identify some of the common species. If you are starting out on your own, have been on an introductory course or are a naturalist with an interest in lichens but without specialist knowledge, this book is for you! We want you to be as wowed as we are by lichens. They are ubiquitous and fascinating windows onto the amazing adaptability and complexity of life on Earth, and can be enjoyed anywhere – from bus stops, parks and pavements to mountaintops and remote rainforests.

In Britain and Ireland, there are more than 2,000 species of lichens. This book highlights a set of species that beginners can identify with nothing more than a ×10 magnifying lens and a little practice. Most of these will be the larger macrolichens, the so-called leafy (foliose) and bushy (fruticose) types. However, since more than three-quarters of the species in Britain and Ireland are small microlichens with a crust-like growth form, we can't ignore them! Not all of these are minute, and certainly not all are difficult to identify. Indeed, some of our most colourful and conspicuous lichens form crusts (crustose).

We offer several alternative **Pathways** to identifying lichens with this book (pp. 44, 64, 85), but first we want to inspire you. We begin with some background information about lichen biology and ecology: what they do out there in the world, and how humans and lichens get along (or don't). Next, we want you to be able to approach lichens in a way that is comfortable and straightforward. The bulk of this guide is formed of **Species Descriptions**, organised according to growth form and colour, with corresponding tabs in the page margins. We also separately introduce some important habitats and their common lichens, as well as groups of related

Colourful but tiny lichens on a concrete sewer in Edinburgh.

The mountains are full of rich-textured and colourful crustose lichens, and some can be identified even when you are just starting out, like this Alpine Bloodspot.

species that you can learn to recognise easily and the features that you need to be familiar with to identify them.

We cover just over 200 species, which have been selected from a set of the most commonly recorded lichens, filtered by ease of identification. A handful of these have been included because of their distinctiveness combined with their ease of identification, and because they are representative of particularly interesting habitats – even if they aren't necessarily common across all of Britain and Ireland.

We are really fortunate in Britain and Ireland to have great identification books for people who already know something about lichens, with excellent resources for intermediate to expert lichenologists (see **Resources and Next Steps** and **Further Reading**, pp. 239–43). In this guide, we have included both scientific names and the common names that you may encounter, mostly coinciding with the names used on iNaturalist (see p. 241). This consistency in naming will allow you to look up the same lichens in these other texts and resources.

Once you have achieved familiarity with some of the common species, there is a lifetime's pleasure to be gained from exploring lichens all year round and almost anywhere at all – and you can start right away! We know this because we've heard the little gasps of awe when someone looks at a lichen through a magnifying lens for the first time.

THE BASICS: LICHENS 101

WHAT IS A LICHEN?

Lichens are fungi, and the bulk of the lichen body – called the **thallus** (plural: **thalli**) – is constructed by a fungus. Fungi tend to grow outwards in ever-increasing rings, being made of incredibly fine hair-like threads growing from their tips. Have you ever noticed that little spot of mould on bread or fruit? It is actually a fungus made of these microscopic threads. Lichens also grow from their tips outward, often forming round patches, and are built from the same incredibly fine, hair-like fungal threads, known as **hyphae**. Several different layers within a lichen thallus (see opposite) are made of hyphae – sometimes packed tightly and modified for protection on the upper or lower surfaces, sometimes loosely woven to promote airflow.

However, lichens are not simply fungi alone. Symbiosis or 'living together' is central to the lichen story, because fungi, just like us, need to eat: some fungi eat by decaying and absorbing dead matter, others by forming an association with plant roots for the exchange of food and nutrients, while lichenised fungi get their food by growing tiny gardens inside their thalli. These minute gardens are populations of single-celled green algae or cyanobacteria that make their own food (sugars) by photosynthesis using solar power. These photosynthetic symbiotic partners (the green algae or cyanobacteria) are called **photobionts** and they are often found in a layer just below the upper surface of the thallus, displayed to the sun. These green algal or cyanobacterial cells are enmeshed by the fungal hyphae, allowing transfer of photosynthetic sugars to the fungus as its energy source.

The symbiotic relationship between the fungus and its photobionts determines the beautiful and sometimes complex structure of the lichen **thallus**, which houses and protects the photobionts from the dangers of UV radiation and from the constant threat of being eaten by grazers that range in size from mites to reindeer. You can think of lichen thalli as living greenhouses built by the fungi, in which to farm their photobionts for providing food. The complex structures that lichens often produce also support a wide variety of other life, including bacteria and other fungi that live on and in the thallus. Some of these offer important vitamins for growth, among other roles, but most of the mineral nutrients needed by a lichen probably come from minute quantities of dust that settle on their surfaces, or are absorbed in solution from rainwater.

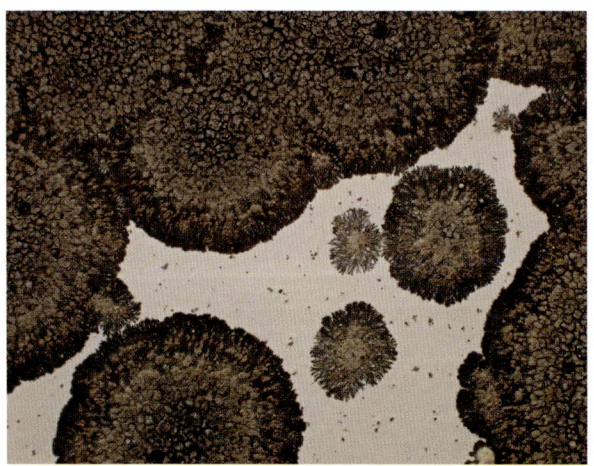

Lichens, like all fungi, begin life as microscopic spores or tiny fragments and grow outwards radially, in ever-advancing fronts, often forming circular growths. This is *Buellia aethalea*, growing on a smooth painted surface.

BACKGROUND IMAGE: Cross-section of *Parmelia sulcata*, at 400× magnification. The complexity of this lichen is clear when you look at a cross-section under the microscope. All the different layers are specialised for different purposes, and all are made of fungal tissues, except the thin layer of green, single-celled algae.

OVERLAY: *Parmelia sulcata*, shown 4× life size.

fungus: upper protective layer

algae and fungi: nutrient exchange

fungus: layer for air flow

fungus: lower protective layer

What is a lichen? 7

Humans are relatively big organisms, and we easily forget that most of life is small. Lichens offer a visible interface for us to appreciate the tiny scales at which most of biology operates. Lichens have microscopic stages of their life cycles, yet they also make visible and beautiful thallus structures that offer a window onto the diversity around us.

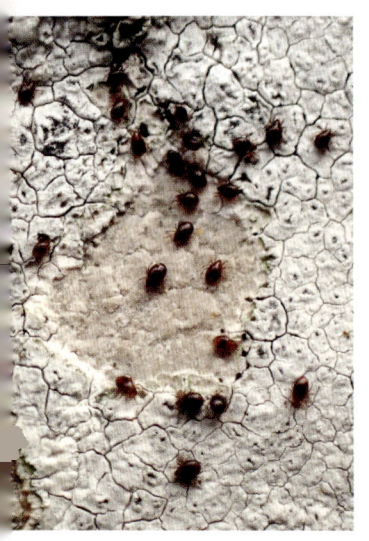

A damaged thallus of *Circinaria calcarea* with orobatid mites.

WHAT DO LICHENS *DO*?

Lichens may be small, but they often have large roles to play in the ecosystems they occupy, from nutrient capture and cycling, soil formation and stabilisation, to shelter for invertebrates, and making vital contributions to our understanding and monitoring of environmental health.

Nutrient cycling

Within the lichen symbiosis, the photobiont turns atmospheric carbon dioxide and water into simple sugars via photosynthesis. This is called primary productivity. In feeding themselves in this way, lichens create food for other organisms too, as they die or get eaten. In some habitats, such as boreal forests and Arctic tundra, lichens can contribute important fractions of all the primary productivity in the ecosystem.

Most lichens are small, and so are many of the creatures that feed on them, including mites, each about a millimetre

8   The basics: lichens 101

Lichens make food – that is, simple sugars – from carbon dioxide and water, using photosynthesis (just like plants). Hence, they are primary producers, forming the base of an entire food chain and being eaten by other creatures. They are rich in carbohydrates.

As a food source, lichens may be best known for supporting the vast herds of caribou in North America or the reindeer in Europe and Asia, and even in Scotland the Cairngorm reindeer herd eats lichens too. These large herbivores survive harsh winters by consuming mat-forming *Cladonia* called Reindeer Lichens, but many other lichens are consumed as well.

A member of the Cairngorm reindeer herd demonstrating that reindeer not only eat lichens on the ground or growing on trees but even on rocks.

What do lichens *do*?

The springtail *Orchesella cincta* on *Xanthoria parietina*.

Telltale signs of mollusc grazing on a *Lecanora* lichen.

in size, springtails and other insects like caterpillars and beetles. Molluscs such as snails and slugs favour some lichens over others – and they are often selective in eating the protein-rich parts of lichens including the spore-producing structures, leaving characteristic scrape-marks where they have been feeding.

Some lichens are also able to add nitrogen into ecosystems, taking inert nitrogen gas from the air and fixing it into their tissues, which then become available for other creatures to access once the lichens die or get eaten.

About 10% of all lichens can fix nitrogen in this way, gaining access to this growth-limiting nutrient. This is important especially in wet habitats such as rainforests, where this nutrient is easily washed away.

A snail grazing on *Lobaria pulmonaria*, a nitrogen-fixing lichen.

A community of nitrogen-fixing lichens in Scotland's temperate rainforest.

The blackish curls are the lobes of the nitrogen-fixing Jelly Lichen *Enchylium* growing among mosses on limestone.

Many lichens, like this *Stereocaulon vesuvianum*, grow directly on bare rock.

Primary colonisation

How do lichens survive on bare rock, and how do they attach?

Lichens are among the first forms of visible life to take hold on newly exposed surfaces. This is possible because of the unique combination in which fungus and photobiont form a self-sustaining unit. In essence, lichens are fundamentally similar to independent ecosystems, with a photosynthetic primary producer (photobiont) and consumer (fungi) linked in the same thallus. The fungi can absorb nutrients from the air while their hyphae can attach by growing into microscopic gaps within rocks – breaking down minerals and enhancing weathering – while the photobiont provides the supply of food as sugars.

This mutualistic relationship explains why lichens are excellent early colonisers of land surfaces, including bare rock. They can grow on and anchor into even notoriously hard rocks such as granite or quartzite, beginning the slow process of soil formation by physical and chemical alteration of the rock surface.

Over long time periods, chemical interactions between lichens and rocks result in the development of soils, as the rocks break down into smaller particles, lichens and other small organisms die and decay, dust is trapped, and organic matter accumulates. This process can be observed in the mountains, where lichen growth is

LEFT: Chemicals from this lichen interact with the rock, releasing iron, which then rusts when exposed to the air; many lichens have rusty colours.

RIGHT: These young Map and Rock Lichens (*Rhizocarpon geographicum* and *Fuscidea cyathoides*) are growing on hard quartzite rock. Combine the unique biology of fungi with a photosynthetic symbiosis, and lichens have superpowers for primary colonisation.

The microscopic threadlike hyphae that build lichens, like this *Diploschistes*, can grow in and among soil particles, binding them together and preventing erosion of soil surfaces.

gradually replaced by mosses, ferns and flowering plants, sometimes resulting in conspicuous mats of vegetation atop boulders. The same process occurs where there has been glacial retreat, landslips, or even through quarrying or other post-industrial effects where lichens can play an important role in re-vegetating bare surfaces and supporting biodiversity.

When growing directly on soils, the fungal hyphae of lichens can also bind particles together, stabilising the soil and building crust-like blankets that retain moisture and nutrients in desert and alpine ecosystems.

This ability to grow on newly exposed surfaces and to build and stabilise soils, which eventually supports other plant and animal life, makes lichens an essential part of primary colonisation and biological succession.

Shelter and camouflage

Lichens are more than just food for grazers. Large, leafy lichens can be good places for small creatures to shelter, and many invertebrates have specialised camouflage coloration to hide against a background of lichens. The Scarce Merveille du Jour (*Moma alpium*) is a perfect example.

But it isn't only small creatures that use lichens or employ lichen-like coloration as camouflage. One of the most surprising examples comes from a group of herbaceous plants in the genus *Amorphophallus*, found across the African and Asian tropics. Each plant produces a single large leaf, some with stalks the size of tree trunks, patterned with lichen-like splotches – presumably to convince would-be herbivores that those tree-trunk-like stems are woody (being covered in lichens) and so difficult or impossible to eat.

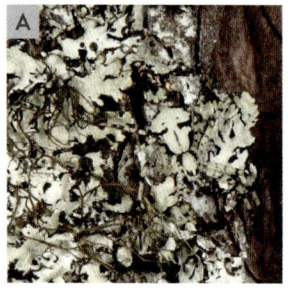

In Britain and Ireland, a number of creatures have conspicuous lichen-like patterning, including the Lichen Running Spider (*Philodromus margaritatus*, **A**) and the Scarce Merveille du Jour (*Moma alpium*) moth (**B**).

This *Amorphophallus* growing in the glasshouses at the Royal Botanic Garden Edinburgh has patches of pale coloration that appear remarkably similar to lichens. Does this fool herbivores into thinking this herbaceous soft tissue is actually woody and therefore inedible? Some scientists think so.

Indicators of environmental health

Lichens help humans to monitor the health of our local environments. Scientists use the very specific needs of individual lichen species to document changes, including in climate, air pollution and habitat quality. This is achievable because, especially in Europe, the environmental tolerances of individual lichen species have been well characterised. For example, studies of sensitive lichens preserved on bark in pre-industrial, timber-framed buildings of England have given us a window into changes through time, documenting substantial losses in lichen diversity with industrialisation and as air pollution became prevalent. Similarly, from studying the distribution of different lichens in wooded landscapes, it was discovered that lichens can be used as indicators for 'habitat continuity', the stability of habitats over long periods of time in a landscape. Lichens – and other small but sensitive organisms like mosses and non-lichenised fungi – are still used today to monitor air pollution and to rate the quality of woodlands for supporting rare and threatened British and Irish biodiversity.

Some of the pale patches on these willow wattles are pollution-sensitive lichens. This panel was removed from a house in Essex built in the pre-industrial seventeenth century. At the time the willow wattles were cut, they were home to a collection of species no longer found in this part of lowland England.

Lichens like these *Cladonia* mats in Culbin Forest in Aberdeenshire can cover huge areas in boreal forests.

WHERE DO WE FIND LICHENS?

Lichens can be encountered on every continent on Earth. They form the dominant ground cover across huge areas of northern forests, Arctic and alpine tundra, and even some deserts, as well as on coastal rocks, on stabilised sand dune systems and growing on trees in forests worldwide. They are conspicuous parts of our British and Irish landscapes, from our polluted city centres to pristine mountain summits.

Habitats and microhabitats

Each lichen species has its own requirements. For instance, consider the climate that a lichen experiences just in terms of moisture and temperature; these factors can affect lichens on at least two different scales. Some lichens thrive in drier and warmer areas to the east of Britain and Ireland, and others in milder, moister areas farther west. These major climatic differences influence the distributions of lichens on a large scale. But nested within these climatic differences there are also important and much smaller-scale contrasts that impact lichens: their **microhabitats**. These microhabitats include any local factors that determine where a species might grow, including shading, exposure to rain or other moisture sources (dew or humidity), and the structure and chemistry of the surface that the lichen is growing on, its **substrate**. A substrate might be anything from damp rotten acidic wood to calcium-rich hard limestone to smooth plastic or painted surfaces, as long as there is a spot where a lichen can establish.

Almost anything can be colonised by lichens given enough time, or enough dispersal from nearby lichens. Notice that the letters on this sign – just a tiny nook to settle in – is where lichens got their start.

From the point of view of a lichen (i.e. measured only in a few centimetres!) a single tree can support many very different microhabitats, from the youngest twigs with lots of light and airflow, to the oldest mossy bases of trunks, where moisture levels are higher, or to the rain-sheltered dry sides of old rough-barked trunks.

So, when we think about exploring lichen diversity, it is important to consider different scales – from major climate types to distinct habitats like seashores versus woodlands, to the tiny microhabitats that support individual species. A single ancient woodland can be home to hundreds of different lichens, partly because of the presence of different tree species, but also due to the variability in microhabitats including smooth and rough bark, deadwood, rocks, soil and streams, each of which might have its own set of special lichens. Even rocks have their own microhabitats, from shaded, rain-protected underhangs to exposed, sunny tops. A favourite bird-perching spot on the highest point of a rock can create an ideal microhabitat for some nutrient-loving lichens, which can tolerate or possibly even rely on the chemical conditions created by the presence of bird droppings.

Another dimension that is easy for us humans to overlook is the fourth dimension: time. The longer a habitat persists undisturbed, the more chance there is for microhabitats to develop and to be colonised by specialist species, and this can take hundreds of years. Some specialist lichens in the Caledonian pinewoods of Scotland, for example, require trees that are centuries old to die, lose their bark and remain standing before their own particular microhabitat develops. Studies in the conifer forests of western North America suggest that the development of new microhabitats is still happening there after 700 years. However, other lichens are 'weedy', able to colonise, reproduce and disperse elsewhere within only a few years, including species that grow on young twigs.

You can see two very different microhabitats side by side on this old oak: the dry, furrowed bark with pale grey *Lecanactis abietina* and the brilliant yellow *Chrysothrix candelaris* in the crevices, and the horizontal mossy branch sporting *Evernia*, *Parmelia* and *Usnea*.

You can often spot bird-perch rocks by their different-coloured tops, usually yellow or orange – and by the droppings. This boulder is topped by nutrient-loving lichens such as *Candelariella vitellina*.

> **DID YOU KNOW?**
>
> On a global level, Britain and Ireland are recognised for having unique concentrations of special lichen habitats, for which we have an international conservation responsibility. Most of these concentrations are characterised by having maintained their microhabitats in the same location over long periods of time (centuries, or even millennia): temperate rainforests, ancient parkland trees and Caledonian pinewoods are three of our most important woodland types, but oceanic mountains, limestone pavements and freshwater habitats also support special assemblages of lichens on their soils and rocks.

Clean air

In addition to climate and microhabitat, differences in air quality are important for understanding where you might find lichens. Do all lichens need clean air to grow? The short answer is 'No, but most do'. For instance, during the 1970s, a famous study of the air quality across Britain was undertaken through a citizen science project, where a thousand schoolchildren returned surveys designed to use lichens as air-quality indicators. By tasking children to tally up the different types of lichens in their local surroundings, scientists were able to show that large areas around big cities had almost no lichens at all and even larger areas had only one or two species.

At the time of the schoolchildren's survey, acid rain was a common side-effect of unregulated sulphur dioxide pollution, and it was a dangerous chemical for humans and lichens alike. The dangers of polluted air were recognised with the passing of Clean Air acts starting in 1956, and these have been effective, reducing acid rain in most areas of Britain and Ireland, and allowing lichens sensitive to acid rain to recover.

However, in the wake of reductions in sulphur dioxide and acidity, in many places the environment that has replaced acid rain is not clean air, but a new nitrogen-enriched atmosphere, leading to hypertrophication, or the overabundance of nutrients. With excess nitrogen, a set of sensitive species is lost. A common sight as a consequence of this is found in intensive farmland areas or along busy roadsides where nitrogen compounds are abundant as a result of concentrated

Lecanora conizaeoides is an acid-tolerant species that was ubiquitous throughout large parts of England and Wales when acid rain was a common occurrence. Today, it is difficult to find.

In the 1970s, the Royal Botanic Garden in Edinburgh had only two leafy lichens and no bushy species, and now it is awash with dozens of different lichens, including some highly pollution-sensitive species.

Xanthoria parietina is usually a sign of ammonia enrichment, in this case surely from canine visitors.

animal husbandry, chemical fertilisers or engine combustion; here, the nitrogen-loving lichen *Xanthoria parietina* is dominant.

The impacts of too much nitrogen are not limited to agricultural or roadside settings, although these can be the worst affected. Unfortunately, nitrogen compounds can be transported in air over long distances and have been shown to be causing a large-scale transformation even of remote environments including the Scottish Highlands, where mountain vegetation is responding to the new elevated nutrient status, with grasses gaining dominance and outcompeting low-growing lichens.

As with the response to acid rain, legislation is in place to ameliorate nitrogen pollution. The trend across large parts of Europe is mostly towards stabilised or reduced emissions, promising recovery. However, the challenging cocktail of changing air pollution still requires conservation attention for protecting sensitive lichens across all our important habitats.

Nutrients may be so high that no lichens can survive and surfaces are covered only in fluffy green algae.

THE BEAUTIFUL INTRICACY OF LICHENS

FUNGI AND PHOTOBIONTS

Remember that lichens are the symbiosis between a fungus and a photobiont. The vast majority of lichens are formed by a group of fungi called the cup fungi, so named because they bear their fungal spores in saucer- or cup-shaped sexual structures (commonly known as **fruiting bodies** or '**fruits**'). This group of fungi is formally called the Ascomycetes, named because their

Lichens come in a rainbow of colours and a smorgasbord of shapes, all because of the intricate relationships between a diverse range of fungi and photobionts – and their need to reproduce.

Ascomycete fungi are called cup fungi because many conspicuous species have open cup- or saucer-shaped fruiting bodies. This cut-away of a fruiting body shows the thin orange layer inside the cup where the microscopic sacs of spores stand packed together, with their openings pointing upwards.

spores are produced in microscopic tube-like sacs (*ascus* is Greek for sac). A much smaller number of fungi that form lichens are members of the Basidiomycetes, the mushroom-forming fungi and relatives; they form their spores on mushrooms, clubs or brackets. In Britain and Ireland, only a handful of lichens are formed by Basidiomycete fungi, including *Lichenomphalia ericetorum*. The mushroom is the fruiting body of the fungus, which produces the spores, but the fungus is still symbiotic with green algae as its food source. The lichen thallus is at the base of the mushroom, made up of the tiny dark green spheres, each of which is a few algae, wrapped in fungal hyphae.

Photobionts in lichens are evolutionarily diverse, representing widely divergent lineages of life. However, about 90% of lichens have a green-algal photobiont, with a single lineage

This mushroom is the spore-bearing part of the lichen *Lichenomphalia ericetorum*, produced by the dark green granular lichen thallus growing over a *Sphagnum* moss.

The cut edge of a *Parmelia* at 30×, showing the thin, grass-green algal layer above a white layer of loosely woven fungal hyphae.

involved in nearly half of all known lichens (*Trebouxia*, a green alga). Photobionts are not as well studied as fungi, but it appears that there are about 20 times fewer photobiont species compared with fungal species involved in the lichen symbiosis. Most fungi that form lichens partner with one or a few green algae that tend to be closely related, and the low – but still important – diversity of photobionts at least partially explains the distinct distributions and microhabitat preferences of different lichens.

> **ACTION**
>
> **TRY THIS** – You can do your own experiment to see the photobiont layer in a lichen: take a pale greyish-green or grey lichen from a twig, tear a lobe off, and then look torn-edge-on, just below the upper surface with a hand lens at 10× magnification. You should just be able to see a very thin, bright grass-green photobiont layer above a thicker cottony-white layer.

Cyanolichens

The other 10% of lichens are **cyanolichens**, in which the photobionts are photosynthetic bacteria called cyanobacteria, hinting at the slightly different colour of the photobiont cells, which are often bluish green, and appearing darker when wet. Cyanobacteria not only make simple sugars for food by photosynthesis, but they can also fix atmospheric nitrogen into organic forms used for growth.

Cyanolichens tend to live in microhabitats with abundant liquid water: in rainforests, in mossy microhabitats, or even on limestone where water pools briefly. These lichens tend to be darker grey and/or brown colours, and this helps identify them in the field.

In Britain and Ireland, there are a few dozen lichen species that associate with both green algae *and* cyanobacteria in different parts of the same thallus. In these species the cyanobacteria are housed in special organs where nitrogen fixation can take place (see *Placopsis* and *Stereocaulon*, pp. 202 and 116, respectively). Perhaps surprisingly, a very few lichens are found with either *or* both green algae or cyanobacteria, and they can look quite different depending on which of the alternative photobionts is present (see *Sticta canariensis* images, opposite).

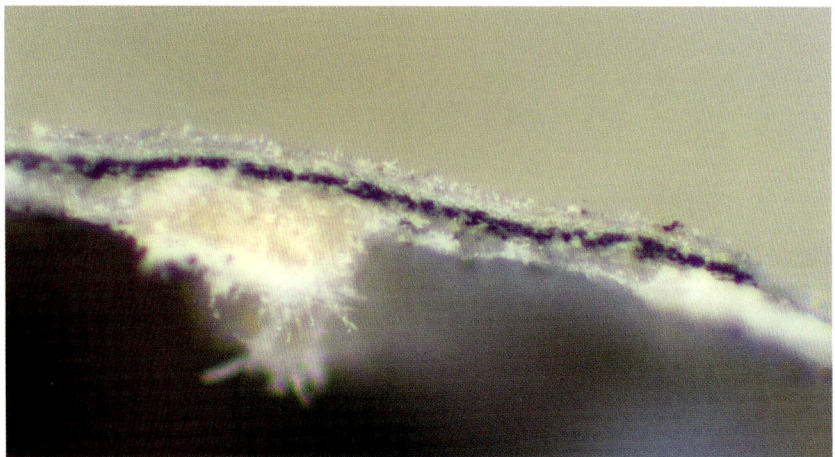

Peltigera has a dark-looking photobiont layer made of cyanobacteria; view of torn edge at 30×.

Lobaria with green algae (right) and *Peltigera* with cyanobacteria (left). When wetted, the outer layers of lichens become more transparent, so green algal lichens look greener, while cyanolichens are darker grey, brown or even blackish colours.

These images show the same fungus *Sticta canariensis* with different photobionts creating different thalli – the thallus with greyish lobes contains cyanobacteria, while the greenish thallus contains green algae. Sometimes, even a single thallus with different-coloured lobes can be found.

LICHEN NAMES

What's in a name?

Lichens are an important contributor to Earth's total biodiversity, numbering in the tens of thousands of species worldwide. Each 'lichen species' is in fact a different fungal species, because the fungus lends its name to the whole lichen. Lichens are, therefore, placed squarely within the evolutionary tree of life of all fungi.

In Britain and Ireland, there are about 2,000 lichen species known, with more being discovered every year. That's one tenth of the world's known lichen diversity and 40% of European lichen diversity. This rich lichen diversity is directly owing to the dramatic variation in our climate, geology and topography, and the variety of habitats and microhabitats.

Plenty of other life is recognised within the thalli of lichens, starting of course with the photobionts. The green algae or cyanobacteria within lichens have their own names. However, we can't identify these microscopic photobionts without specialist laboratory techniques like microscopy and culturing or DNA sequencing. Consequently, the lichen itself – containing green algae or cyanobacteria, or both, as well as fungi – is referred to just by the fungal name, as the fungus makes up most of the lichen tissues, and it probably largely drives the overall structure and shape of the lichen. Even where a single fungus makes very different-looking lichens with a green algal photobiont versus a cyanobacterial photobiont, these very different-looking lichens will still be referred to by the same single fungal name, as in the case of *Sticta canariensis* on p. 21. Lichens also contain a rich complement of non-photosynthetic bacteria and other, non-lichen fungi, and there is a highly active field of research investigating the diversity, distribution and roles of these different partners in the biology of lichens.

Changing names

Biologists organise living things into nested groups to help understand their features and relationships. Species are often recognisable as distinct from one another just by looking at them. A collection of closely related species is brought together by scientists into a grouping referred to as a **genus** (plural: **genera**) – just as dogs and wolves are in the genus *Canis*, Pixie Cup and Reindeer Lichens are in the genus *Cladonia* (p. 66).

Over time, the boundaries of these genera and other larger groupings like families change, as new information about their relatedness comes to light. This is because scientists who study the diversity of life try to reflect genetic and evolutionary relationships in their classifications, and each new piece of technology leads to new insights. Over recent decades, the use of DNA to classify organisms including lichens has resulted in shifts in our traditional classifications, which were previously based on physical similarity alone.

In this guide, we adopt a practical approach to learning. We offer common names for individual species and for groups of lichens that share similar features. For instance, many of the orange crustose lichens were until recently grouped into a single large genus containing very many species called *Caloplaca*, but these have since been split into more, smaller, genera. Occasionally, the new classifications separate similar-looking species into different genera because of their evolutionary relationships. However, in this book we continue to find value in teaching the larger and still-recognisable groups; in this example, those yellow or orange crust-forming lichens with a characteristic set of features in common, known as the Firedots, are the '*Caloplaca*' group.

Likewise, the large genus *Parmelia* has been split and split again into different smaller genera, but you can start to recognise them as members of the '*Parmelia*' group (meaning 'like *Parmelia*'), and we refer to these as the Shield Lichens.

Because the understanding of relatedness among species is still in flux, the names of individual species must necessarily change to accommodate this new understanding. For each species in the Species Description pages we use the new, current, name and sometimes show an older name in parentheses, which will help you 1) connect with other resources and 2) learn to assemble species into useful larger groups. For instance, *Variospora flavescens* was until recently in the large genus *Caloplaca*, but is now placed into a smaller group that is named as the genus *Variospora*. In the Species Description pages you'll see both names, including the traditional name in parentheses like this: e.g. *Variospora flavescens* (*Caloplaca flavescens*).

Common names

Few lichens are mentioned in everyday language in most cultures, and certainly few have ever had English names in common usage. The Gaelic word 'crotal' or 'crottle', referring to many different lichens used in dyeing, or 'litmus' and 'orchil' for the purple-dye-producing *Rocella* species, are a couple of examples from Britain and Ireland's traditional and early industrial heritage. We recognise that scientific names can feel intimidating, but over the years many authors have offered different English names. Most of the common names in this guide correspond to those proposed by the authors of the first richly illustrated *Lichens of North America*, who chose descriptive names that have often been adopted by the online lichen recording community on iNaturalist; in a few cases, we chose to show names used in regional guides produced in Britain and Ireland. Occasionally in the notes, we include quirky nicknames that provide useful memory aids for growth form or structural features.

A number of lichens have been used in dyeing and had common names like crottle, korkir or orchil.

Ramalina fastigiata is called Dotted Ribbon in iNaturalist, but the possibly more memorable Shrek's Ears or Fanfare of Trumpets may be better descriptors for some people. For any lichen that you want to remember, choose a nickname that works for you!

LICHEN IDENTIFICATION

WHERE TO BEGIN – A FEW QUESTIONS

First things first. There are three questions you should always try to answer to help you identify a lichen. Think about these as different scales of observation, from habitat and microhabitat, to whole specimen, to small physical features you will need to observe closely:

1. Where is it growing? By that we mean, where in the country, and on what – knowing where exactly you have observed something will help you identify it.
2. What is its growth form?
3. What are its key features, including any reproductive structures?

RIGHT: In lichen-rich places like this, don't get overwhelmed; first, just consider all the different combinations of growth forms and colours, then look for how each lichen reproduces. If you do this, you'll probably be able to sort out all the species even if you don't know their names.

TWO IMPORTANT THINGS TO NOTE

Can I identify absolutely everything?

Simply put, no. First of all, lichens have to start somewhere, and many will be too young to be identified, even by expert lichenologists. **A handy rule is: if it isn't reproducing, walk away!** See pp. 31–36 for reproductive structures. Also, remember that there can be quite a wide range of variation even within a single species, but some lichens you find will be outside the normal range of variation; those oddball specimens may defy you.

Also:
Do look around; try to find well-developed examples of a species before you try to identify it.
Do start with the larger lichens forming bushy tufts or leafy structures, and wait awhile before you tackle the crust-forming lichens.
Do check the troubleshooting section at the back for some common problems (pp. 234–38).

How do I look at a lichen? Hand lens versus by eye

Different features of lichens should be observed in different ways. In this guide, we've mostly selected lichens that are easy to identify just by eye – some with a bit of practice! But having the ability to see just a bit more always helps. For some of the smaller lichens, and some of the features of most others, a hand lens will help: magnification of 10 times is standard, and we indicate the need to use a hand lens with 🔍 in the text. It's worth spending a little extra money to purchase a lens with a light source – you'll be glad you did!

1. WHERE IS IT GROWING?

If you start to think about where something is growing, you will begin to learn what species you are likely to encounter in different situations, and that will narrow your job to a smaller set of possibilities - so it's well worth considering properly. For something small, like a lichen, 'where' encompasses both a **habitat** (e.g. urban park or mountain heath or temperate rainforest) and a much smaller-scale **microhabitat**, or the location within the habitat where the lichen actually grows.

Take a look at our habitat pages (pp. 44–63) for thumbnail images of conspicuous lichens in selected habitats. Starting to take note of these larger habitat differences is an important part of learning which lichens grow where.

When you zoom right in to the lichen-sized microhabitat, this will help you narrow your identification choices even further. Is the lichen growing on rock, soil, mosses, bark, wood? Many lichens are quite choosy at the small scale of exactly what they are growing on, whether that is smooth, young bark or old, rough bark, or different types of rocks (see pp. 58–61) and exposures, be it sunny or shady, in a run-off track of water or a dry underhang.

Taking note of this information helps you build knowledge of the individual lichen with its habitat and microhabitat preferences, and it can help create confidence in your identification based on the next two questions. Step closer to the lichen to answer these...

Make sure you have the hand lens almost touching your eyelashes (or glasses), and then bring the lichen up to the lens OR lean your head in close towards the lichen until it is in focus.

2. WHAT IS ITS GROWTH FORM?

The growth form can be recognised easily just by looking at the lichen in its natural home, often unaided by magnification, except for the smallest species.

There are three main growth forms, and, to interpret them, it helps to understand the internal layering that most lichens have. The outer surface of most lichens is called the **cortex**. It is usually made from thick-walled, close-packed fungal threads (hyphae), which offer protection and an absorptive surface. In cross-sections under the microscope, the angular outlines of the thick-walled cells can be seen in a blocky matrix (see p. 7 and illustration, opposite). Directly under the cortex is the **photobiont layer**, on display to the sun to allow photosynthesis to occur. Below the photobiont layer are the loosely woven fungal hyphae of the **medulla**. This layer has large air spaces between hyphae to allow gas exchange – carbon dioxide in, and oxygen out – for photosynthesis. See the illustrations for each of the growth forms on the following pages to understand how these layers are structured.

Fruticose

Fruticose lichens (pp. 87–124) are bushy, hairlike or have stalks, which can be upright, sprawling on the ground or dangling down, usually with a distinct three-dimensional form. The photobiont layer is typically found on all sides of the branches or stalks of fruticose lichens, making the lichen look the same in colour and texture, any which way you turn it.

Fruticose lichens are often bushy or tufted, like this *Usnea wasmuthii*.

Many fruticose lichens have a holdfast, which is a strong single attachment point. Branches arise from holdfasts or as individual stalks. You may already have seen a couple of the more obvious fruticose lichens, like *Usnea* and *Cladonia*, even if you don't know their names.

There are many different *Cladonia* species with stalks, which may be pointed, richly branched or tipped with cups, like this *Cladonia coccifera*.

Hair lichens like this *Bryoria fuscescens* are also fruticose; like all fruticose lichens, their branches look the same all the way around without a distinct top and bottom.

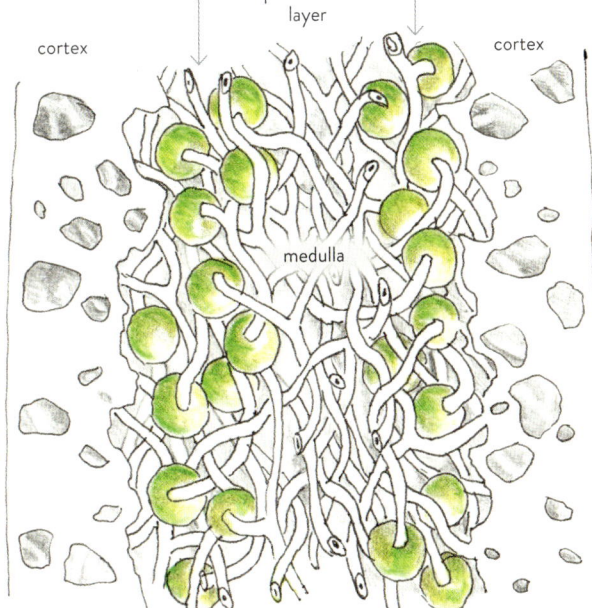

Formally speaking, fruticose lichens have a photobiont layer that runs all the way around the inside of each branch of the lichen more or less completely, so the branches usually look the same no matter which way you turn them.

Foliose

Foliose lichens (pp. 125-84) are made of flattened leaf-like parts, and they spread mainly in two dimensions. They differ on the upper and lower surfaces, and the edges of their lobes can be peeled from what they are growing on fairly easily, for example with a fingernail or the tip of a penknife. They can be very large and floppy or tiny and closely pressed to bark or rock.

Foliose lichens like this *Peltigera hymenina* have leafy lobes, which look different on their upper and lower surfaces. This is because the photobionts are located near the upper surface only, exposed to the sun.

The flattened, repeating parts of foliose lichens are called **lobes**. They have widely varying sizes that are measured in terms of their individual width, away from the branching points and away from the tips – lobes of different species can be characteristically very small, less than a millimetre wide, or rather large – even lettuce-sized (see p. 39 for how to measure lobes).

One of the main reasons that the upper and lower surfaces of foliose lichens are different is the distribution of the photobiont layer, which occurs only near the upper surface, exposed to the sun. The lower surface is variable depending on the species, from smooth and black with a cortex, to white and cottony without a cortex, and with or without root-like extensions

Lobes come in all shapes and sizes, from more than a centimetre wide as in *Peltigera* (above), a few millimetres as in *Parmelia* (far left) or mostly less than a millimetre wide such as in *Physcia* (left).

Foliose lichens have a photobiont layer just below the upper surface, so the lobes look different on the top compared with the bottom.

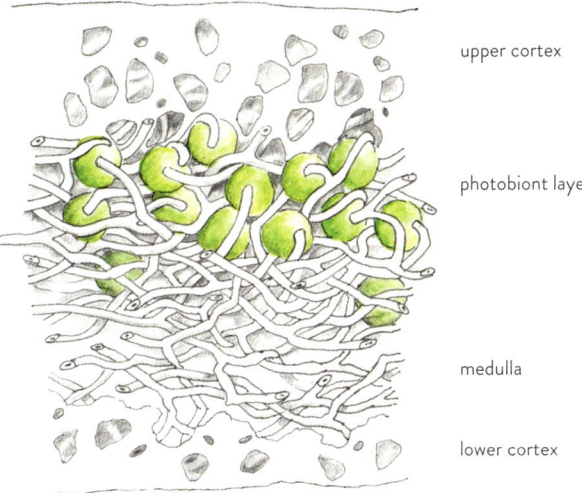

upper cortex

photobiont layer

medulla

lower cortex

called **rhizines** (see p. 37), which attach the lobes to what they are growing on. Internally, there is often a thick, white loosely woven medulla layer within the thallus, below the photobiont layer.

Squamulose

A sub-type of foliose lichens is the **squamulose** lichens. These species are formed of small scale-like structures, called **squamules**. The scale-like leafy squamules are usually crowded together, or can be smaller and separate, and they each attach individually to the surface they are growing on. They are not attached across their entire lower surfaces to the substrate (what they are growing on) and they have different upper and lower surfaces; you'll find these at the end of the foliose section in the Species Description pages (pp. 182–84).

Many *Cladonia* species start life as squamules, and some are made mainly of squamules for their entire life cycle, like this *Cladonia subcervicornis*.

2. What is its growth form?

Crustose

Crustose lichens (pp. 185–233), often referred to as 'crusts', are mostly flat and crust-forming, and they may look like a stain or a blotch of paint from a distance. They are attached across their entire lower surfaces to their substrate by their threadlike fungal hyphae; remember that these hyphae are so microscopically small that they grow between individual bark cells, grains of rock etc., meaning that they cannot be peeled away with a fingernail or the tip of a penknife. In order to collect crustose lichens, you must normally collect the substrate they are growing on to keep the lichen intact. The thallus itself may be thin to thick, smooth, warted, wrinkled, powdery, granular or even cracked like dried mud. Sometimes they grow predominantly within the bark or rock itself, with only spore-bearing structures visible on the outside. Sometimes the growing edge is slightly lobed or is of a different colour.

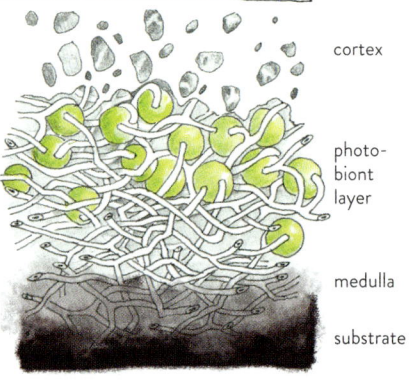

ABOVE: Crustose lichens like this *Rhizocarpon* follow the contours of the surfaces on which they are growing.
ABOVE RIGHT: Crustose lichens are attached to their substrate across their entire lower surface.
RIGHT: A lobed crustose lichen, *Variospora flavescens*.

Lobed crustose

Watch out, there are a few crustose lichens that look foliose around their margin, because of a rosette-like or lobed growing edge, but these are so closely attached that they can't be removed with a penknife or fingernail; this makes them crustose lichens – **You will see them labelled as 'lobed crustose' in the Species Description pages (pp. 188–91, 198–202).**

ACTION

TRY THIS – Head outside and see if you can spot all three main growth forms. Since fruticose lichens tend to be the most sensitive to air pollution you might not be able to find these unless you visit somewhere away from busy roads, heavy industry, city centres or intensive farming, which are all sources of air pollution. When you find what might be a foliose or fruticose lichen, check to see if it has a distinct upper and lower surface.

3. WHAT ARE ITS KEY FEATURES?

Two types of feature are important to study: those involving reproduction, and those which are vegetative, or non-reproductive, parts of the thallus.

TOP TIP Quick reminders of the most important features are shown on the back flap of the book. On a lichen you want to identify, try to remember to observe what the reproductive structures are, what shapes they form, *and* where they are found on the thallus!

Reproductive structures

Reproductive structures provide a number of useful features for identification, and when you start to look at lichens with a hand lens, you will be able to see them. Reproduction happens in two main ways in lichens:

1. Vegetative reproduction via the dispersal of various types of lichen fragments, by which we mean the fungus *plus* photobiont. There are various methods promoting a breaking off of thallus pieces, in which the fungus plus its photobiont disperse together. This is a type of asexual reproduction, since there is no genetic recombination, but merely clonal fragments dispersing from the original parent thallus.

2. Reproduction via production and dispersal of tiny fungal spores that are produced in fruiting bodies of various types. These spores can be either sexual, including a genetic recombination (sexual) step, or asexual, in which spores are clones of the parent lichen fungus. Sexual spores are likely to have novel combinations of genetic material that allow for lichens to thrive in new settings.

Consider the fungal life cycle. A fungal spore, tiny and microscopic, may land on a suitable spot and begin growth as a network of microscopic thread-like hyphae. These hyphae branch and spread until they come into contact with a suitable photobiont, the preferred green-algal or cyanobacterial partner depending on the lichen species. Once the symbiotic relationship is established, the fungus grows a typical lichen thallus to house its photobiont, and to build up enough resources to grow to maturity and then reproduce all over again. In spore-producing species, the lichen fungus must re-associate with a suitable photobiont. Although it may seem unlikely that a microscopic fungal spore may by chance encounter and re-lichenise with a suitable microscopic green algal or cyanobacterial colony, all evidence suggests that it is a common feature in lichen biology; this separation of the partners also facilitates potential switching between different photobionts, which may benefit the lichen in different environmental settings.

Vegetative fragments tend to be larger than spores and might not disperse as far, but create new thalli more easily, as they carry the photobiont along with the fungus. Different species tend to reproduce mainly by one strategy (sexual spores) or the other (vegetative fragments), though many if not most are probably able to use both methods.

TOP TIP Lichens grow from their tips, so the youngest parts are the edges, and the oldest parts are in the middle. When looking for specific features, especially reproductive structures, use your hand lens to look at the youngest parts first, then work your way to the older parts to follow their development.

The lichen life cycle, showing sexual (left) and asexual (right) modes of reproduction.

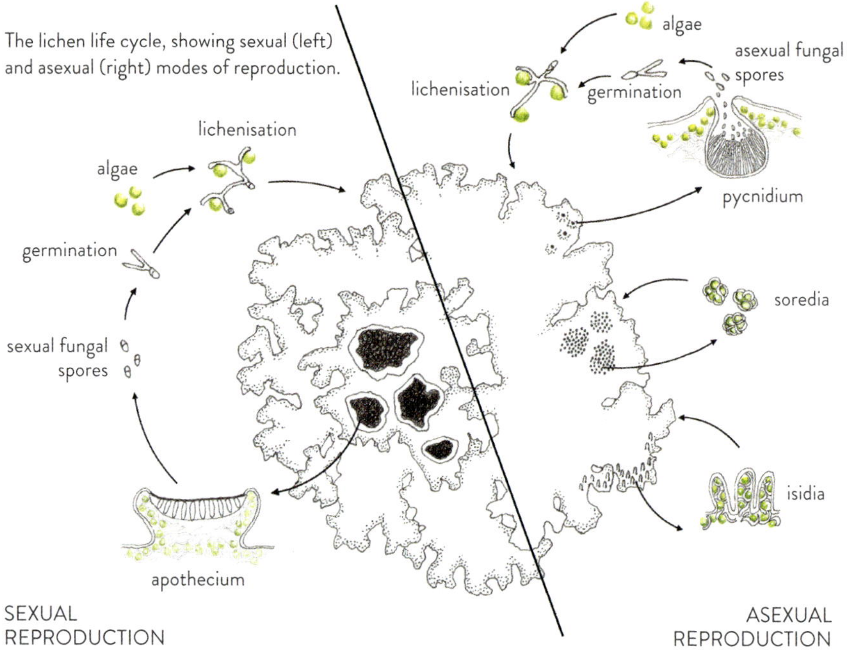

SEXUAL REPRODUCTION

ASEXUAL REPRODUCTION

Vegetative reproduction in detail: fragmentation, isidia and soredia

Many lichens may have a relatively easy time of reproduction, by just breaking a fragment off their thallus, which can grow into a new lichen that is genetically identical to the parent, already containing both the fungus and photobiont. For example, Reindeer Lichens (pp. 113–14) fragment easily when trampled on if they are dry, and these smaller genetically identical pieces can then disperse and re-grow. In the grand scheme of things, these broken fragments tend to be relatively large and don't travel very far from the parent thallus. Other lichens make various specialised pre-packaged vegetative fragments in different shapes and sizes.

Isidia (singular: isidium) is a general term for pre-packaged lichen fragments arising directly from and continuous with the upper surface of the parent lichen. Because they originate as outgrowths

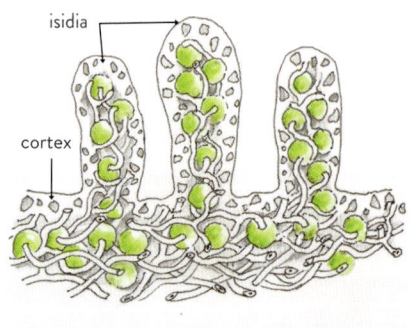

Isidia arise from the upper cortex layer. Cross-section of *Melanohalea exasperatula* with isidia in side view (above), and stylised illustration of isidia, showing fungal and photobiont tissues within (right, not to scale).

of the upper surface, the cortex, they have the same texture as that upper surface – if you aren't sure, look with your hand lens at how they develop, from the youngest ones near the edges of lobes to the older ones nearer the centre of the lichen. Isidia are often raised like tiny fine fingers, and sometimes richly branched, or sometimes rounded and inflated or flattened, depending on the lichen species.

Soredia (singular: soredium) are even smaller than isidia, and appear as tiny powdery or crumbly grains, each being made up of an envelope of fungal threads encasing and entangling a few photobiont cells. A single soredium is only visible with a hand lens, but often they are grouped together. Lichens with soredia are often referred to as **sorediate**. Individual soredia themselves are different sizes depending on the lichen species, from fine and floury to coarse and granular (like granulated sugar). Soredia arise from the inner parts of the lichen, the medulla, so look for a break in the upper surface to find them.

Soredia arise from the inner layer of the lichen, the medulla, via breaks in the surface. Above: surface view of clusters of soredia arising from cracks in the cortex, and (below) stylised illustration (not to scale) showing cross-section of thallus producing soredia.

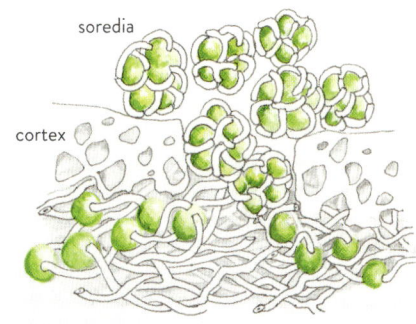

Soralia are groups of soredia in distinct clusters, which can be found in different, but

Vegetative, asexual reproduction by fragmentation in lichens. Clockwise from upper left: **fragments** of branches break and disperse in fruticose *Cladonia portentosa*; **isidia** on the foliose lobes of *Parmelia saxatilis*; mounded clusters of coarse **soredia** in crustose *Lepra amara*; and oval patches of fine **soredia** on edges of branches of fruticose *Ramalina farinacea*.

predictable, places on lichen thalli, depending on the species. This positioning can be helpful for identification. Soralia are also very variable in size and shape, from tiny and dot-like, to large and mounded in profile, crescent-shaped (as seen on *Parmotrema perlatum*, p. 171) or even lip-shaped (*Physcia tenella*, p. 180).

Sexual reproduction in detail: apothecia, perithecia and mushrooms

The other mode of reproduction – concerning only the fungus – is sexual reproduction. Sexual structures are different across different branches of the fungal tree of life. As a reminder, most lichen fungi are in a huge and diverse group called the Ascomycetes (cup fungi; p. 19), which make several different kinds of sexual structures. Fortunately, lichenologists like to keep things simple (for once!) and use only two names for these, depending on whether the fertile surface is exposed or covered as it develops.

Apothecia are fruiting bodies normally with exposed fertile surfaces which look like tiny saucers or jam tarts. The 'jam' is the exposed spore-bearing layer, full of tiny sacs of microscopic spores (and other threadlike cells to keep the sacs upright so that spores can be ejected from their tips). The jam is called the **disc** and can be different colours depending on the lichen species. The jam tart crust is called the **margin** (sometimes called the **rim**), and it is usually slightly raised. The margin is what makes the 'cups' of all the so-called cup fungi – it is a shallow dish-like container that contains the sexual spore-bearing surface. We will use the term **margin** for this structure throughout this book.

Margins of apothecia are useful to observe for identification. Sometimes, the margins of apothecia look the same as the thallus, and so are called 'thalline margins' or 'lecanorine margins';

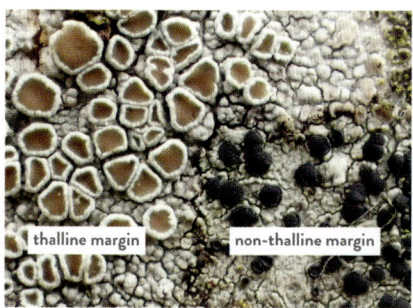

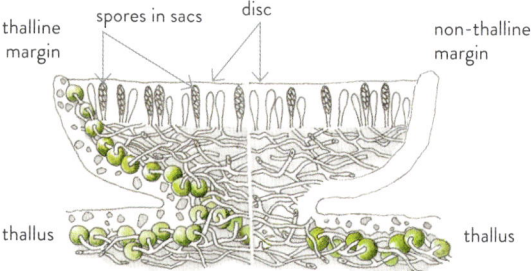

ABOVE: Jam tarts are a useful way to explain apothecia, with the crust representing the cup-like margin, often raised at the edge, and the jam representing the spore-bearing surface.
ABOVE RIGHT: Two types of margins in apothecia: thalline margin (left, on *Lecanora*) or non-thalline margin (right, on *Lecidella*).
RIGHT: A cross-section diagram of apothecia showing the algae in the margin on the left and without algae in the margin on the right.

this is because the margin contains the photobiont within it just as the thallus does (see p. 82, Rim Lichens). However, apothecia don't always have this thalline margin. In the *Lecidea* group (p. 83), the margin is often coloured the same as the disc, and is unlike the thallus because it has no photobiont within. This is called a 'non-thalline margin' or 'lecideine margin', more like a wine-gum than a jam tart.

Some apothecia are elongate, like little scribbles of script. These are called **lirellae** (p. 84).

Perithecia are fruiting bodies that have covered fertile surfaces, which look like a chocolate-covered marshmallow teacake. In this analogy, the chocolate is the cup-shaped container. If you imagine that you could take tiny tongs and stretch the entire margin of an apothecium up and close it together like a flask to almost fully enclose the jam (disc), you would have formed a perithecium. The margin forms an enclosed flask (the chocolate on the tea cake), open only at a tiny pore at the top for spores to be ejected. Inside are the sacs of spores (where the marshmallow is!). Perithecia are found in a few different crustose lichen groups and tend to be inconspicuous.

One more structure associated with reproduction sometimes causes a bit of confusion with perithecia, so we'll discuss that here. On the surface of some lichens are very regular tiny black dots, well under a millimetre in size, and under the hand lens may be seen to have a tiny pore at the top. These are a type of **pycnidia** (see life cycle above and image p. 32), tiny structures that contain thousands of even tinier asexual spores, some of which are probably implicated in mating (i.e. equivalent to sperm, bringing new genetic material to a separate receptive cell). The asexual spores may also be able to form a new lichen thallus as a mode of asexual reproduction, if they contact a suitable photobiont.

There are also a very few lichens that belong to the Basidiomycetes (mushroom-

A cut-away apothecium of *Tephromela atra*, showing the disc (black), and the thalline margin with green algae, which looks just the same as the whitish thallus – also broken to show the algal layer within.

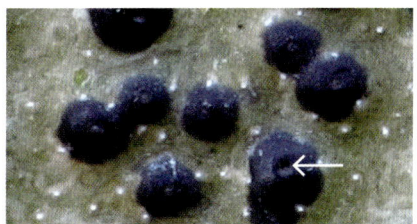

Tea cakes are a useful way to explain perithecia, with the chocolate coating representing the cup-like margin completely covering the spore-bearing surface – the marshmallow. Tiny pores for spore release (lower photo) are visible on tips of perithecia (arrow), which are otherwise entirely closed.

3. What are its key features?

forming fungi), and these can form all sorts of fruiting bodies – in the tropics from fans and clubs to toadstool-shaped mushrooms. Here in Britain and Ireland, our most common Basidiomycete lichens produce mushrooms. These are small and beautiful gilled mushrooms which you would never have suspected to be formed from lichens, until you look at their bases to see tiny green lichenised granules or scales.

Spore-producing structures in lichens.

Non-reproductive thallus features

The thallus of a lichen may or may not include some other more or less obvious features to observe in identification, and we will focus on a few essentials here while others will be introduced where needed in the Species Descriptions (pp. 90–233) and can be found in the glossary.

Jelly Lichen

Most lichens have a distinct photobiont layer; you can scratch the lichen to look for a green or dark layer just beneath the upper surface. But some lichens aren't layered at all – the so-called **Jelly Lichens** have fungal and photobiont cells intermixed throughout their thalli within a matrix of gelatinous material. These lichens can be rather thin and papery when dry, but the gelatinous matrix swells when wet – sometimes markedly so. They have cyanobacterial photobionts and are usually dark brown, dark green to blackish; they can be any growth form, but most often are foliose. See the Baker's Dozen page about them on p. 78.

Jelly Lichens are dark in colour and can swell markedly when wet. Their cyanobacterial photobionts produce a jelly-like matrix and the fungal threads are mixed with photobionts within the jelly, not forming distinct layers.

Hair-like structures – rhizines, cilia and tomentum

Rhizines are primarily anchoring structures and they *grow from the lower surface of the thallus*. They can be small or large, branched in various ways or not, pigmented or not. Sometimes they protrude from underneath the edge of the thallus, but where they do occur, they are always attached on the lower surface. If you are unsure, remove a lobe and check the point of origin with your hand lens.

Cilia are hair-like structures that *grow directly from the edges of the thallus*. They can be confused with protruding rhizines, but, with your hand lens, check the point of origin to help you decide.

Tomentum is a *fine and even lawn of short hairs* or *fine felty covering*, best seen with a hand

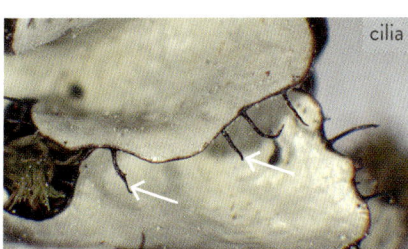
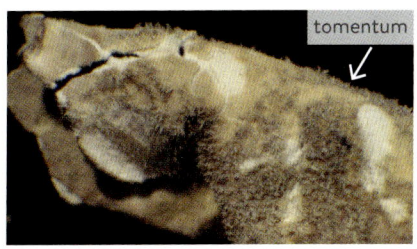

3. What are its key features?

lens. It can be found on the upper surface of some *Peltigera* species or the lower surfaces of some cyanolichens including species in the genera *Lobaria*, *Lobarina*, *Pseudocyphellaria* and *Sticta*.

Surface features

Pruina is a very *fine crystalline coating*, found on the upper surface of some lichens, sometimes even on discs of apothecia. The lichen may look whitish by eye, and under a hand lens the pruina will appear as a fine dusting like icing sugar.

Pseudocyphellae appear as scattered *white or paler spots, lines or cracks, where the inner layer of the lichen is visible* and exposed to the air. Their role is to allow some exchange between the inside and outside of the lichen, and so there will be a textural contrast with the surrounding cortex surface when examining them closely with a hand lens.

Areoles. In crustose lichens, the thallus is often formed as small islands or develops by cracking into smaller pieces as it grows. These *smaller portions of a crustose thallus* are called **areoles**, and the way they form and develop can be useful to observe since they can be characteristic for different species.

Size

Lichens are special among the fungi because they are visible year-round. This makes them among the best-known and most obvious types of fungi. However, lichens exist at a different and smaller scale than humans, so getting to grips with what to expect in terms of size can be a challenge. When a novice learns to identify birds, there is often a standard set of questions: smaller

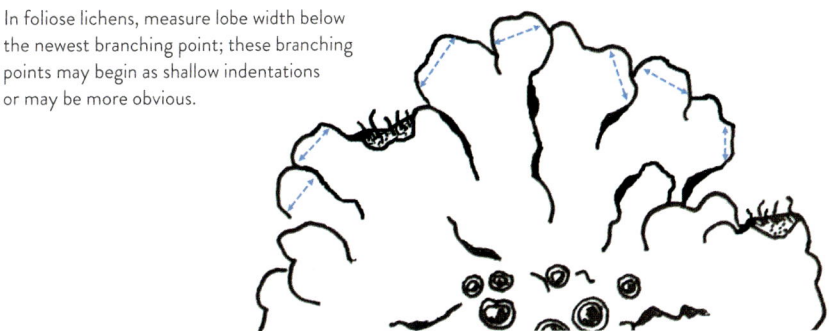

In foliose lichens, measure lobe width below the newest branching point; these branching points may begin as shallow indentations or may be more obvious.

than a blackbird? Bigger than a crow? Does it have the bill of a seed-eater like a sparrow or a bill for eating insects, like a robin? Lichens challenge our expectations for describing even something like size because they can just keep growing, and often several smaller individuals will grow together to form quite extensive patches. Lichens have their habits, but be prepared for lots of exceptions in their overall thallus size. Because of this, we encourage a focus on lobe sizes, measured as width, away from growing tips and away from branching points. Of course, these will vary, too, but usually less so than the overall thallus.

ACTION Points of reference

TRY THIS – We suggest that you try to find examples of different-sized lichens just to get started. Here are a few points of reference, shown to scale.

Head to a tree with lots of grey lichens, maybe in a park, and see if you can see blue-grey or green-grey lichens of different sizes. Find larger lobed lichens with white markings, which could be dot-like or irregular in size and shape – pseudocyphellae – on their lobe tips. These are members of the Shield Lichens group, *Parmelia* and relatives. What about tiny pale grey lichens with lobes about a millimetre or two in width? These are in the Rosettes, Frosts and Fringes group, *Physcia* and relatives. Now, can you find a yellowish-orange leafy lichen with darker orange apothecia? This may be Sunburst Lichen, *Xanthoria parietina* (**1**). Look at the lobe sizes; these are fairly regular, even though the full size of the lichen thalli will vary.

In this book, Shield Lichens like *Punctelia* (**2**) are medium sized – with lobes 2–5mm wide, while Rosette, Frost and Fringe Lichens like *Physcia* (**3**) and *Physconia* (**4**) are narrow lobed, with lobes usually less than 2mm in width.

Colour

Just as with size, it is important to reorient your expectations a little with lichen colours. Colours are products of the basic machinery of a lichen's biology – often produced as pigments in order to regulate light, protecting themselves and their photobionts from damaging UV. This means that the colours of thalli can vary within species or even within individuals depending on exposure to sun, but also with wetness or overall thallus health. We have a few guidelines about colours.

- Lichen colours refer to the colour of the lichen when it is dry.
- Watch out for conspicuous colour changes when wet. This is because the upper surface of the lichen – its upper cortex – becomes translucent and the photobiont layer becomes easier to see. We indicate these in the Species Descriptions (pp. 90–233) with a water drop symbol 💧. Green algal lichens turn greener, while cyanolichens will become darker – brownish, bluish or blackish.

Physconia grisea is greyish when dry and green when wet.

In this *Xanthoria parietina*, the lower parts of the thallus are yellowish because they protrude out of the hedge just slightly more and thus receive more direct light, while the more shaded parts are a less typical greyish green.

- Be aware of exposure. Lichens exposed to bright sun may have different colours from the same species in the shade.
- Know that lichens get sick and die, too. Unhealthy lichens change colour, sometimes dramatically; for example, *Parmelia* species often turn bright red directly under caustic bird droppings (the basis for some common chemical tests you can find out about as an intermediate learner). Some common lichen parasites are brightly coloured as well; see Troubleshooting (p. 235).

We divide lichens into major colour families:

If your lichen is **unmistakably yellow or orange**, at least in part, count yourself lucky: you have just narrowed the potential number of species dramatically! Yellow and yellowish-orange are distinctively bright and deep colours that you would be unlikely to confuse, and there are relatively few groups of lichens that have these colours: check *Xanthoria*, *Firedots*, *Candelaria*, *Candelariella* and *Teloschistes*.

Some clothiers have adopted 'lichen' as a colour, since **pale whitish-grey to greenish-grey** hues are common across many lichens. *Parmelia*, *Punctelia*, *Physcia* and *Platismatia* are all pale greenish-grey; however, the shadier and moister the conditions, the greener these lichens will be. In this guide, we also include in this category lichens that have a yellowish-green pigment (usnic acid) like *Flavoparmelia* or some *Xanthoparmelia* species. We also include very pale to whitish lichens as these will often become greenish or greyish in certain conditions.

Relatively few lichens are really **bright green** unless they are wet. However, those that are include *Lobaria*, *Ricasolia*, a few *Peltigera* species, and *Solorina*. Others like *Physconia*, *Phaeophyscia*, *Anaptychia*, some *Melanelixia* and crustose lichens like *Baeomyces rufus* can be green, but tend to be more muted and not as bright.

Khaki, or **greenish brown to brown or blackish** colours are common, both in green algal Camouflage Lichens and also among cyanolichens. The Camouflage Lichens (*Melanelixia*, *Melanohalea*, *Xanthoparmelia*) can be easily overlooked, blending into bark or rocks. They tend to be darker in higher light situations, becoming greener in the shade. In wetter environments, where cyanolichens become commoner, Moon Lichens (*Sticta*) and Pelt Lichens (*Peltigera*) can often be brown when dry, sometimes darkening almost to blackish when wet. Dark grey to brown or black lichens are often cyanolichens.

Red or rusty-red or rusty-orange crustose lichens can be eye-catching, but many are beyond the scope of this book. Check the 'Wine-gum Lichens – *Lecidea* and other black dots on rocks' page in the Baker's Dozen, and the rusty-red crustose lichens on pp. 83 and 233.

Although the rule is to describe colours when dry, in reality, lichens that live in wet places are often wet, so we show some of these species wet as well.

HOW TO USE THIS FIELD GUIDE

PATHWAYS TO LICHEN LEARNING

In this book we are offering **three main pathways** for your lichen journey – three ways to discover and identify lichens presented in the Species Descriptions. You can use any pathway that you choose; each one is intended to help you find your lichen by narrowing the choices of lichens you may be trying to identify. From each of these pathways, you will be referred to page numbers in the Species Descriptions to find out more. There are Species Description pages for 133 lichens; a further 104 similar species are mentioned. On the side of each page there are growth form icons, and lichens within each main growth form are organised into broad colour categories. Each Species Description includes important features for ID, habitat information, notes on similar species, and range maps, so be sure to check these.

Before embarking on any of the above pathways, we recommend reading the introductory material (pp. 24–41) and learning a few descriptive lichen words that will crop up throughout this guide – **Don't forget that the front and back flaps have reminders of basic lichen terminology!**

1. Lichens by habitat
Habitat descriptions (pp. 44–63) give you a brief introduction and thumbnail images for common or iconic lichen species you can expect to find in some habitats that are good for lichen hunting – for example in a woodland, on a coastal rocky shore or a mountain heath. Look for a thumbnail image that matches your lichen best, and then check against the corresponding Species Description in the **Lichens by form and colour** section.

2. Important lichen groups: the Baker's Dozen
We have provided a Baker's Dozen (pp. 64–84) of some of the most important and conspicuous lichen groups you are likely to find. These groups and the species they contain can be useful as anchor points or references, when getting to know how to interpret colours and sizes, and creating expectations about the types of features that need to be recognised for identifying lichens. On these pages, we also offer suggestions of what to focus on for identifying the individual species within these groups.

3. Lichens by form and colour
The main part of this book is made up of Species Description pages, organised by growth form and colour. First, lichens are arranged by the main growth forms, fruticose, foliose and crustose. Within the three main growth forms, sets of lichens are broken into the main colour families (orange and yellow, brown and black, pale 'lichen' colours, as described above) and further arranged by shared features like sizes and textures of branches (fruticose lichens) or lobes (foliose lichens), or type of reproduction (crustose lichens).

You can use the Quick Guides (pp. 88, 126 and 186) for each growth form to help narrow your choices to the lichen species that match best with what you want to identify. Using the Quick Guides will get you to a set of species that share similar features and are presented together in the Species Description pages.

SPECIES DESCRIPTION PAGES

Each Species Description page gives detailed information of individual lichen species and any similar species, to aid your lichen identification.

The main lichen species described on this page

The bold icon will guide you to sections on FRUTICOSE, FOLIOSE or CRUSTOSE lichens

The form and colour set found in PATHWAY 3: LICHENS BY FORM AND COLOUR (p. 85)

A lavender grey bar here shows which BAKER'S DOZEN group a species belongs to, if any (p. 64)

A text section containing the following information:
Scientific name – may include a recent older name in parentheses
Common name/s
How to spot – The 'gestalt' or search image of this species: i.e. what to look out for from a metre or two away
Description – The main features, reproduction type/s and diagnostic characteristics
Where – The substrates this lichen can be found growing on and the wider habitat it can be found in. Specialised habitats may also be highlighted here.
Notes – Extra notes of interest on this species or group of species

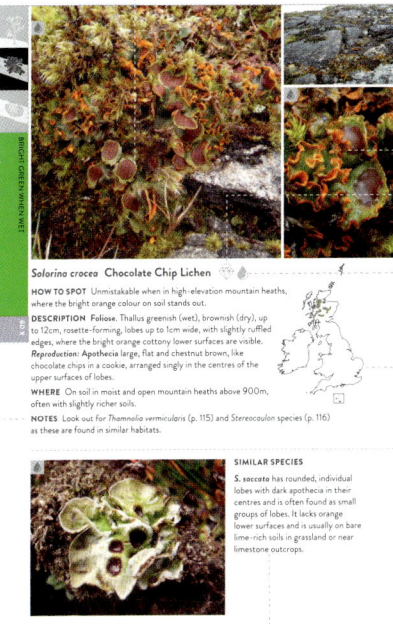

A visual example of the type of habitat in which you may come across this species. **Arrows** have sometimes been added to help you spot the lichen

Key features to aid identification

A rare or threatened species in part or all of its range. Please enjoy just looking – DO NOT COLLECT

Lichen changes colour strongly when wet

The distribution across Britain and Ireland. These maps are based on information from the British Lichen Society databases 1980–2023.

Brief descriptions of similar-looking species and how to tell them apart. Species names in **bold** type have corresponding pictures here. Occasionally, these will be species found in the same habitat, but not similar in form and colour.

Finally, it's important to understand that identification can be thought of in terms of probability – and sometimes, even for experts, confidence is not 100%. This book should help you name or get pretty close to a name for the commonest and most conspicuous lichens in Britain and Ireland. Although it's not always possible to identify every specimen, the Troubleshooting section (p. 234) should help if you find yourself stuck.

WE HOPE YOU ENJOY YOUR LICHEN JOURNEY!

THE PATHWAYS

LICHENS BY HABITAT

Some people prefer to explore their own local patch, or they may have a habitat-based perspective when it comes to learning about the diversity around them. In this section of the book, we offer a short description of how to look for lichens, particularly with reference to different types of places – habitats – you might explore. Thumbnail images of **common** or **iconic species** for a handful of different habitats are provided to give you a menu of lichens you may encounter. Some species will be found in more than one habitat, and others are specialists for a single habitat, so be sure to check the Species Description pages.

Thinking like a lichen

But first, we invite you to start thinking about *how* to look for lichens. It's all too easy to walk up to a tree, rock or wall, and look at it from a convenient human height. However, you will be missing a lot if you limit yourself to that approach. Don't forget, humans are quite big, and lichens are small – and so lichens respond to tiny differences in position, exposure, moisture, chemistry and so on, which define their **microhabitat**. Different lichen species prefer different microhabitats, so the more you tune in to the scale of lichen microhabitats, the more you will discover.

Just as an example, consider rocks: for humans, rocks can be good to sit on for a short rest in the sunshine and off the damp ground. Take a moment to imagine the differences you would experience if, as a lichen, you lived your whole life in the rock's microhabitats: the tops may be bird-perches, but these along with the sides, bases, crevices, sheer faces and underhung corners all provide very different microhabitats in moisture and exposure alone. Remember, within any given habitat, there can be a multitude of microhabitats.

The habitats

To give you a starting point, we offer a few notable habitats and outline some associated lichens you'll find described in detail in the Species Descriptions. Although most of the lichens described in this book are common species, we also include a few less common, maybe even rare, but very distinctive lichens, that you might go hunting for on a lichen safari. For each habitat, the thumbnail images are arranged first by growth form, and within that by their other similarities following their order in the Species Descriptions. These lists are just tasters – a handful of species for each habitat, and the species are not always exclusive to the habitat they are listed in.

CHOOSE YOUR HABITAT

- TREES AND WOODLANDS p. 46
    - Deciduous woods and wayside trees p. 48
    - Pinewoods and acid-barked trees p. 50
    - Temperate rainforest p. 52
- CITIES AND THE BUILT ENVIRONMENT p. 54
- MOUNTAINS, MOORLANDS AND HEATH p. 56
- SILICA-RICH ROCKS p. 58
- LIME-RICH ROCKS p. 60
- COASTAL AND SEASHORE ROCKS p. 62

A boulder like this is not the same all over from the point of view of a lichen. Soil gathers on the top, water courses down the mossy front, and there is a dry underhang on the left.

TREES AND WOODLANDS

Trees and woodlands provide a hugely important habitat for lichens, from ancient woodlands and old wayside trees in hedgerows and pastures, to street trees in cities and parks. Different species of tree have contrasting bark characteristics, meaning that different lichens can be found on them. The age of the tree, bark textures and acidity status all play important roles – the acid flaky bark of pines is very different from the stable but less acid bark of ash, and the smooth bark of young trees offers very different conditions from deep furrows on older trees. Some bark is naturally nutrient-rich and supports a distinctive set of species; this bark tends to be only mildly acidic to neutral in pH (ash, elm, aspen, willow). The dust from cities and agriculture can also result in an additional level of nutrient enrichment, which supports only a limited set of tolerant species (* in the list on p. 49 and see pp. 54–55).

Lichens do not harm trees, and they are not parasitic like mistletoe. Instead, they only anchor themselves to the outer layers of bark, just as epiphytic orchids and ferns do in the tropics, allowing them to access light and moisture where they can thrive.

Try to consider all the different microhabitats a tree can offer at the centimetre or even millimetre scale, from the mossy bases and shaded furrows to the sunny side of exposed trunks or smooth bark of twigs.

You won't find the same lichens in every woodland. Regionally, woodlands in western Britain and Ireland tend to experience a much wetter climate than those in the east, and this also affects the types of lichen species that may grow there. Air pollution is still affecting lichens on trees in and near large cities and intensive agriculture. Even in areas which may not feel particularly urban, old trees may still be recovering from very poor air quality of decades past. Many lichens also appear to have limited capacity to move around the landscape and can therefore be very abundant in one spot and entirely absent quite nearby! Taken together, all this means that every woodland is different, but here are some of the common lichens you are likely to encounter on different types of trees and in different types of woodland.

THE ART OF LOOKING AT TREES

TOP TIPS

- Examine the tree from many different angles, aspects and heights: the air could be cleaner up in the canopy than down near a road, so more pollution-sensitive lichens might be found higher up.
- Check the branches, twigs, deadwood, crevices, bark injuries, dry sides, moisture tracks, the arches of roots – all affect where different lichens will grow.
- Lichens on twigs and the ends of branches can be very quick to respond to environmental change, so the species you encounter may be very different from those on the trunk. You may also find that lichens on twigs are very tiny and/or very young, missing the features required for identification.
- Remember that different trees have very different characteristics, especially bark chemistry and texture. Birches and conifers such as Scots pine have acid bark; willows, ash and elm have only mildly acidic to neutral bark; and oaks tend to be somewhere in the middle. Young trees have smooth bark and older ones have deeper furrows. Look at a range of tree species, but remember that individual trees can vary, too.
- It may be tempting to examine newly fallen branches to see what is growing in the canopy, but lichens can often become infected with other fungi and die quickly when they fall to the ground. Because fallen lichens may not look typical or may be sick, they can sometimes be difficult to identify, and so this usually isn't the best way to search for lichens when you are starting out.
- Fruit trees can be a fantastic place to start identifying lichens because they tend to have a good diversity of species on them and are at a convenient height for accessing the canopy!

The differences in lichens on different parts of this oak tree are stark. The trunk below the cut limb is protected from rain and is drier, with its particular crustose lichens, each different colour picking out a different microhabitat; the right side with mosses and larger fruticose lichens is exposed to rain.

An ash tree in a city park supports a very different set of lichens from one in the countryside.

Deciduous woods and wayside trees

Wayside trees and woodlands are an excellent place to start your lichen journey. Old trees with only mildly acid to neutral pH bark like ash, elm or aspen are very important habitats for special lichens and are often worthwhile places to stop and explore.

Usnea subfloridana (p. 98)

Cladonia coniocraea (p. 109)

Ramalina farinacea (p. 122)

Evernia prunastri (p. 130) *Physconia grisea** (p. 177) *Physcia tenella** (p. 180)

Melanelixia subaurifera (p. 159) *Parmelia sulcata* (p. 165) *Punctelia subrudecta* (p. 166)

Flavoparmelia caperata (p. 167) *Parmotrema perlatum* (p. 171) *Hypogymnia physodes* (p. 173)

Xanthoria parietina* (p. 139) *Lepra amara* (p. 217) *Lepra pertusa* (p. 221)

* Species found in cities or on nutrient-rich bark, such as ash and willow

Pinewoods and acid-barked trees

Some trees tend to have rather acidic bark, particularly pine, conifers and birches. These trees often exhibit a different community of lichens from those with less acid or nutrient-rich bark, though there may be some overlap because individual trees can vary in their characteristics.

Bryoria fuscescens (p. 97)

Usnea dasopoga (p. 101)

Usnea subfloridana (p. 98)

Cladonia polydactyla (p. 105)

Sphaerophorus globosus (p. 118)

Evernia prunastri (p. 130)

Temperate rainforest

The woodlands of western Britain and Ireland, and in wet ravines farther inland, include examples of temperate rainforest – globally rare habitats with internationally important assemblages of lichens, mosses and liverworts. These habitats are among the most important for lichens in Britain and Ireland. The species that characterise temperate rainforest depend on year-round mild temperatures and consistent moisture – often with rain on two out of every three days. Rowan trees like the one shown above, along with ash, hazels and willows, support the most iconic temperate rainforest lichens, cyanolichens of the so-called Lobarion, including large leafy cyanobacterial lichens like Lungwort, Felt, Moon, Specklebelly and Pelt Lichens.

Some trees with more acidic bark in temperate rainforest support completely different lichens to those found on rowans, ashes and hazels, and Shield Lichens can be abundant, as seen on this alder.

52   The pathways

CITIES AND THE BUILT ENVIRONMENT

Urban and built environments are filled with contrasting structures and surfaces, and as long as the air quality is not too poor, lichens will make a home there. Metal railings, old signs and lampposts, monuments and walls, along with mortar and concrete, all offer microhabitats for common lichens – and of course, don't forget the trees in city parks and gardens. These are great places to start looking at lichens. Most urban areas have relatively high nutrient input from car exhaust and dust, and there is a predictable set of lichen species that thrive in these warm, dry and nutrient-enriched conditions.

Churchyards are also good places for exploration, in part due to the long stability of their structures and surfaces – church buildings, old walls and gravestones – but also because of the wide variety of different rock types present in one place. Granite is a silica-rich and hard rock, while marble is lime-rich and somewhat softer, so the same lichen species will rarely be found on both. Check the habitat pages for Silica-rich rocks and Lime-rich rocks (pp. 58–61) for some ideas about the lichens found on these different types of substrate.

This park bench in Bristol is home to more than ten different lichens, from large leafy species to inconspicuous crusts.

MOUNTAINS, MOORLANDS AND HEATHS

Large parts of north and west Britain and Ireland are covered in mountains, moorlands and heath. These areas tend to have low nutrient inputs, with expanses of heather and rock. Most of the mountains and moorlands are also generically classified as **uplands**, where the climate is cool and moist. Higher-elevation and rocky habitats are typical of the uplands, but the further north and west you travel, the lower the elevation at which you might find typical upland species, including lichens. Even in the **lowlands**, especially further south and east, similar vegetation and lichens can be encountered on nutrient-poor heathlands, so look out for lots of *Cladonia* among the heather. This habitat has a big overlap with the Silica-rich Rocks habitat (p. 58), but **here we focus on the species that live on or near the soil**.

Kneel down in gaps between plants or look for pale yellowish-green or brown patches of *Cladonia* to discover the diversity of lichens growing on soil.

Lichens by habitat

SILICA-RICH ROCKS

Most of the exposed rocks in the British and Irish uplands are relatively high in silica, including granite, gneiss, sandstones and quartzite. Although these rocks are lacking in nutrients, this habitat probably has more lichen diversity than any other. Lichens here are predominantly crustose species, most of which are difficult to identify when just starting out. Below is a handful of species that can be more easily recognised.

The fluorescent Yellow Map Lichen *Rhizocarpon geographicum* mostly grows only on silica-rich rocks; when you see it, start to look out for other typical lichens of silica-rich rocks and build your lichen-hunting confidence.

LIME-RICH ROCKS

In contrast to silica-rich rocks, lime-rich rocks contain a proportion of calcium or magnesium carbonates, and the lichens you can expect to encounter are very different. In fact, lime-rich rocks are sometimes more easily recognised by their lichens than by any of their physical characteristics, as they provide habitat for very bright and distinctive orange, white and black crustose lichens. Limestone areas include the Yorkshire Dales, the Peak District, the Wye Valley on the border of England and Wales, and the Burren in Ireland. These natural areas are fantastic places to explore, but concrete, mortar and certain sandstones also contain carbonates and so have some of the same lichens. You may see these substrates described as calcareous in some texts, but we refer to them as lime-rich, as a reminder that they are relatively nutrient-rich, slightly alkaline and contrasting with silica-rich rocks.

These concrete kerb stones have the distinctive orange, black and white crustose lichens of lime-rich rocks, like the lichens on the dry stone wall at Malham (above).

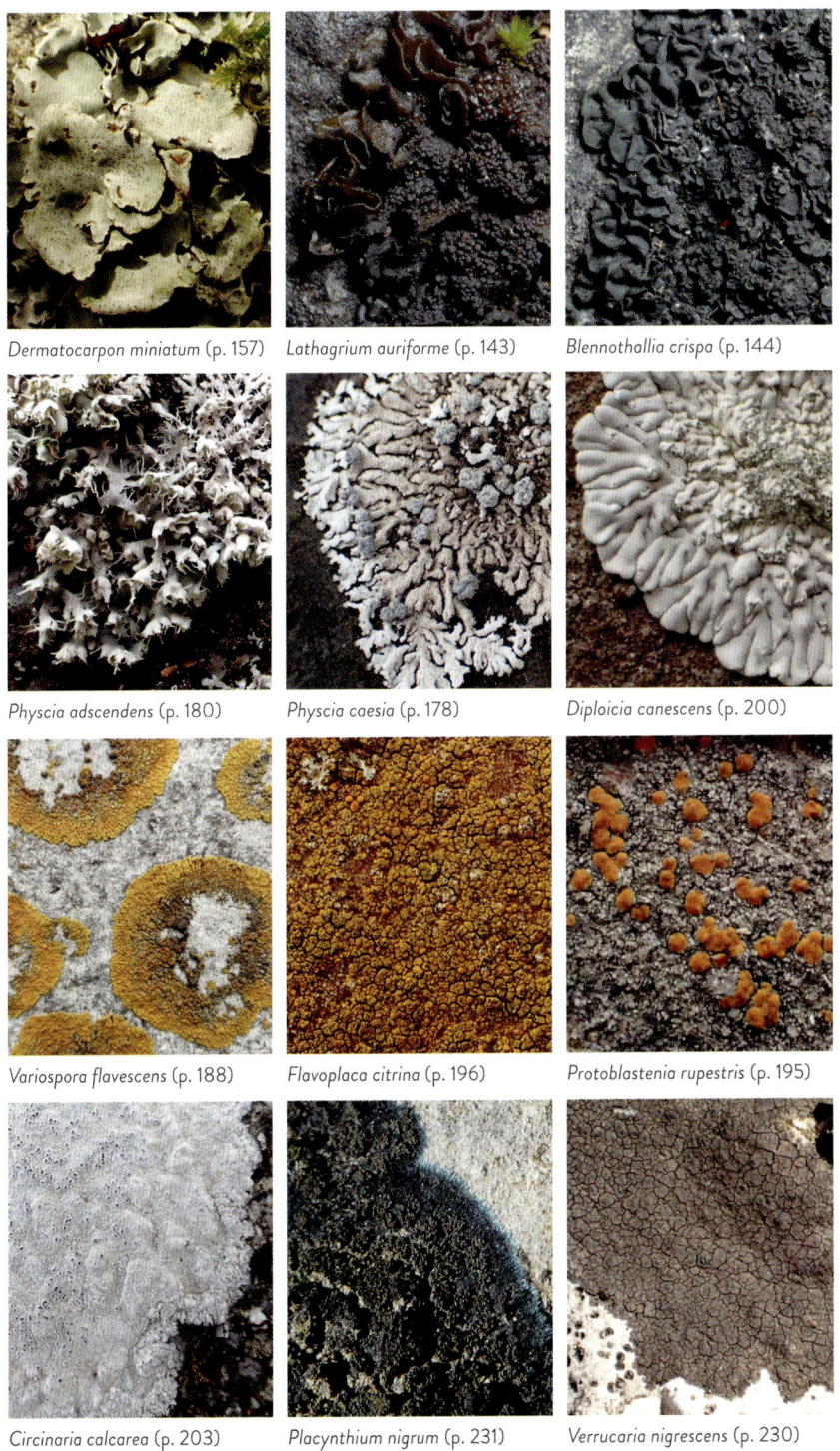

COASTAL AND SEASHORE ROCKS

Coastal rocks, alternately washed by salty seas and exposed to bright sun, can be a hostile environment. Nonetheless, a range of specialised lichens has adapted to survive and even thrive in this challenging habitat. Most coastal lichens grow just above the high-tide line and create the distinctive zones of colour we often see on a rocky seashore – from black in the upper intertidal to orange in the sea-spray zone. The grey zone higher up has the most species, and is dominated by lichens not requiring a seaside vantage. On the lowest tides, you may be lucky enough to spot the greenish waxy circles of the lichen *Wahlenbergiella mucosa* growing alongside barnacles. If you look closely, even the tiny black dots on barnacles themselves are lichens!

We present these species in their respective zones down the seashore, rather than by growth form and colour, as you'll soon learn what to expect to grow where.

Grey Zone – mainly grey lichens, conditions approaching inland habitats.

Orange Zone – splash zone (often mixed with grey or black zones).

Black Zone – upper intertidal, submerged twice a day.

A BAKER'S DOZEN OF IMPORTANT LICHEN GROUPS

When you are getting started with learning to identify lichens, the 13 groups set out below will help you make quick progress. These groups actually cover hundreds of species, with some easy and some not so easy to identify, but if you get your lichen into the right group in the first place, you're already making good headway towards knowing what species it is.

HOW ARE THE LICHENS GROUPED?

As a reminder, biologists organise living things into groups (see p. 22) separated at different levels of relatedness, in a nested hierarchy of classification. The Baker's Dozen groups are often **genera** (singular: **genus**), collections of related species with common characteristics. In practice, genera are often easier to recognise than individual species. Sometimes, where it can be helpful, we also gather groups of genera together. We use the term 'group' here quite flexibly, to refer to a collection of related species within a genus or collection of related genera. In this section we begin with fruticose lichen groups, then move to foliose groups followed by crustose.

Look out for the Baker's Dozen tabs on the Species Description pages so that you can refer back to the information here, for example 'BD1' refers to the Cartilage Lichens in the genus *Ramalina*.

ACTION

TRY THIS – When you visit a new place, first, only look for fruticose lichens. Second, sort these for identification according to similar features – colour, branching structure, attachment, texture – these will often correspond to the genus level. Third, add your observations on differences in their reproductive modes and structures and you'll be sorting into what might correspond to species within those groups. Next, try the same thing for the foliose species. **When you're starting out, just do your best to narrow down a lichen to one of the Baker's Dozen groups below, then you can work to identify the correct genus or species. Begin with the larger fruticose and foliose lichens, then come back later to work on the crusts!**

Ramalina – Cartilage Lichens

Ramalina lichens are **fruticose, pale greenish and strap-shaped**. In Britain and Ireland there are about 15 species, and they mostly form small shrubby thalli, but a few species can become long and dangling or grow as large as a lettuce. There are a few common, easy-to-identify *Ramalina* species that occur on exposed bark, twigs and coastal rocks. Look for multiple flattened branches arising from a single point of attachment – the holdfast. In some species, the branches may become slightly inflated. This group is called the Cartilage Lichens because they have a tough, cartilage-like texture.

| LOOK FOR THESE KEY FEATURES |
- Fruticose growth form
- Shrubby or long-drooping
- Flattened, strap-like branches, pale green all around
- Holdfast at the base, tightly attached to rock or bark

The shapes of branches, types and position of reproductive structures (apothecia or soredia) helps with species identification.

Ramalina fastigiata (**A**) has round, flattened apothecia on its branch tips while *R. fraxinea* (**B**) has apothecia on the branch surfaces and edges. *Ramalina farinacea* (**C**) has fine, floury soredia in oval patches on the edges of its branches, while *R. cuspidata* (**D**) and *R. siliquosa* have warty branch tips with asexual spore-producing structures called pycnidia, which have dot-like openings (🔍).

Cladonia – Powderhorns, Pixie Cups and Reindeer Lichens

Lichens can typically be easily recognised as a *Cladonia* with a little practice. However, some *Cladonia* species can sometimes be very tricky to identify because of wide variation among specimens of the same species growing in different conditions.

In this book we introduce only the commonest and easiest to identify of the more than 40 species of *Cladonia* found in Britain and Ireland. These fruticose lichens can be diverse and abundant in nutrient-poor and acid habitats like heaths and moorlands, old mossy walls and bases of trees, and old fallen logs or stumps.

Most *Cladonia* lichens start life as a handful of leafy scales (**squamules**) on organic matter. Squamules are mostly bright white on their lower surfaces, without a cortex (so no smooth outer layer). Later, the characteristic hollow stalks (called **podetia**) develop. Where present, these stalks identify the thallus as a fruticose lichen. However, some *Cladonia* species mostly form squamules and rarely produce stalks, and these are included in the squamulose section of this guide (p. 182). For most other species of *Cladonia*, you'll need the stalks for identification.

Reproduction occurs by thallus fragmentation, by soredia, or spores. Spores are borne in irregular globular apothecia (above and opposite). In some cases, apothecia can be a striking red colour, but most species produce brown apothecia, and many stalks will be found without apothecia at all.

| TOP TIP | If you don't see red apothecia or yellow-orange stains on the lower surfaces of squamules (🔍), you can assume the apothecia (if present) would be brown.

The scalloped edges of the squamules (**1**) in this *Cladonia humilis* are upturned, showing their white lower surfaces, and the ones on the right are covered in soredia (**2**) fallen from the edges of the wide cups.

LOOK FOR THESE KEY FEATURES

- **Fruticose** or **Squamulose growth form**
- Squamules, usually at the base of hollow stalks in most species
- Apothecia red or brown (🔍), irregular and blob-shaped (globular)

TO IDENTIFY INDIVIDUAL SPECIES, INVESTIGATE
- Stalk shape: branching or not, cups wider or narrower than stalks?
- Stalk texture: smooth, or with squamules or soredia (🔍)?
- Size of soredia: finely floury to granular, like tiny globes (🔍)?
- Distribution and size of soredia or squamules on the stalks (🔍)

Reindeer Lichens can form extensive mats; check colours and how the tips branch. *C. portentosa* (**A**) is greyish and has branching in all directions; *C. arbuscula* (**B**) is yellowish with branch tips bent or tips curved in the same direction. True Reindeer Lichens have no squamules at all.

Pixie Cups have short stalks with wide cups (**3**). The cups may serve to splash soredia out when raindrops fall into them. They may have red or brown apothecia or have none at all; check for soredia and try to observe different textures. *C. coccifera* (**C**) is yellowish without soredia, and *C. chlorophaea* (**D**) is grey to greenish or brown with coarse granular soredia.

A number of *Cladonia* species have **pointed stalks**. Some are unbranched like this *C. coniocraea* (**E**), one of the commonest species. Or they may be much-branched, like this *C. uncialis*, (**F**). *C. uncialis* has a distinct yellowish hue because of the presence of usnic acid, usually best seen in contrast with greyish lichens growing nearby.

Squamules (**1**) can be anywhere on the stalks or just appear on their own. *C. squamosa* (**G**) is named for having lots of squamules, and it often has narrow irregular cups (**4**). Some species consist almost entirely of squamules, like this *C. subcervicornis* (**H**) which is common on upland silica-rich rocks.

Granules (**5**) are rounded, larger soredia, sometimes with a smooth cortex. They can become flattened like these in *C. pocillum* (**I**), or smaller and gritty looking as in *C. floerkeana* (**J**). Granules can be seen individually with a hand lens and contrast with finer soredia, which have a powdery or floury texture, as seen in *C. humilis*, on p. 66.

Usnea – Beard Lichens

In Britain and Ireland, there are more than 20 species of fruticose Beard Lichens in the genus *Usnea*. With their large surface area, they are very sensitive to air quality and are not found where any form of air pollution is high. A few features are all you need to identify a lichen as an *Usnea* no matter where you are in the world: **fruticose, yellowish green, cylindrical branches with a stretchy cord inside.** Hold two ends of a medium-sized branch and gently pull to see the stretchy cord. The yellowish green colour is due to usnic acid, a sunscreening compound produced by the fungal partner.

TOP TIP Different individuals and different species can grow closely together and can even be entwined, so be sure to trace branches back to the holdfast before trying to identify them.

LOOK FOR THESE KEY FEATURES
- Fruticose growth form
- Either shrubby or long-drooping
- Cylindrical branches (mostly round in cross-section)
- Stretchy central cord in all but the smallest branches (**1**)
- A single holdfast (attachment point at the base); check colour of basal parts
- Soredia and/or isidia produced in patches called soralia
- The isidia that arise directly within soralia are an exception to the standard definition of isidia because they are not continuous with the cortex.

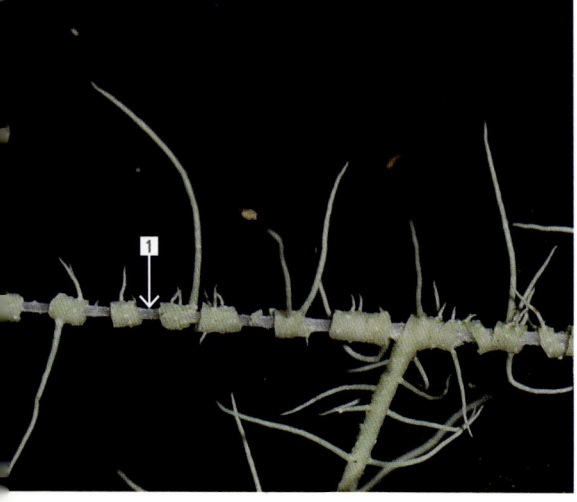

The inner stretchy cord of *Usnea* is visible if you gently pull the ends of a moistened branch.

Form: Thalli can be long, and the term **pendant** is used when thalli are more than two times longer than wide, as for *U. dasopoga* (**A**). They can also be **shrubby**, meaning erect and bushy or somewhere in between, like this *U. wasmuthii* (**B**).

Branching patterns can be distinctive. There are long parallel main branches with rib-like 'fishbone' side branches in *U. dasopoga* (**C**) and *U. rubicunda*. *U. cornuta* has markedly **constricted branch points**, where main branches attach and branching is often at 90-degree angles (**D**, arrows).

Basal parts are often black (**E**), and can have brickwork-like cracks (**arrow**), which is typical of *U. wasmuthii*. In comparison, *U. cornuta* (**F**) has a pale or somewhat brown base.

Soralia are breaks in the cortex where soredia are produced that are usually visible as tiny round patches of different texture on branches (🔍). They and any isidia are essential to examine for the identification of most species. First trace each branch back to the base, to make sure they are all growing from the same holdfast, then check the soralia on different branches, young and old (🔍). Soralia change in size, shape and the presence of isidia along the branches as they mature. (**G**) *U. wasmuthii* young (top) and old (bottom) compared with (**H**) *U. subfloridana* (right) young (top) and old (bottom).

Most *Usnea* species are found without **apothecia**, but nearly all have soralia or isidia, or sometimes both intermixed. Apothecia commonly occur in *U. florida* (**I**), which lacks soralia. Look out for apothecia in other species including *U. subfloridana* in clean air areas of Scotland, and always check for soralia with your hand lens (**J**, arrows).

Parmelia and relatives – Shield Lichens

Moving onto the foliose lichens, one of the first groups to learn is the Shield Lichens, or the '*Parmelia*' group. This is a conspicuous, abundant and diverse group worldwide. Many of the species it contains form rosettes about the size of your palm, some larger, some smaller. A number of common foliose lichens belong here, and at one time they were all considered to be different species in the genus *Parmelia*. Today, they are split into different genera.

Although species and genera in the '*Parmelia*' group may look rather different at first glance, they all share a few features visible with just a 10× hand lens. Many of them are easy to recognise with a little practice.

LOOK FOR THESE KEY FEATURES
- **Foliose growth form**
- Leafy lichens, often rosette-forming
- Lobes mostly 2–5mm wide, sometimes up to 1cm wide
- Green algal photobiont – scratch the surface and you should see a grass-green layer below the cortex (🔍)
- Slightly shiny to glossy cortex on the upper and lower sides (🔍)
- Most have rhizines (hair-like or root-like structures) for attachment on the lower surface

Green-algal lichens including Shield Lichens don't need liquid water to rehydrate and grow – only water vapour, so they will often occur in places with good airflow that dry out frequently. When wet, most turn greener because of their algal partners showing through the upper surface.

Parmelia (**A, B**) has lobes 2–5mm wide, with irregular white lines and cracks (pseudocyphellae) in the upper cortex (🔍), sometimes making a network pattern at the often square-shaped lobe tips. Most species are greenish- to bluish-grey like *P. sulcata* (**A**), but they can be brown to black like *P. omphalodes* (**B**).

Punctelia (**C**, thallus damp) have white dots (pseudocyphellae) on the lobe tips that look a bit like specks of salt (🔍). *Flavoparmelia* (**D**) have rounded yellowish-green lobes, without pseudocyphellae but with wrinkles (🔍). Both of these genera have soredia in Britain and Ireland and are very common in the west and south.

Platismatia (**E**) thalli are often rather large (over 10cm) and loosely attached, with crisped and wrinkled upright lobes with white, brown or black lower surfaces. *Parmotrema* (**F**) are also loosely attached and large, but the upper surfaces of lobes are smooth and even, with dark brown to black lower surfaces.

Xanthoparmelia Rock Shields are found almost exclusively on rock, growing tightly pressed onto the rock surface. Two common species, *X. conspersa* (**G**) and *X. mougeotii*, are yellowish green, and others like *X. verruculifera* (**H**) are yellowish to brown.

Hypogymnia (**I**) and *Menegazzia* (**J**) both have inflated thalli – the lobes are hollow and therefore appear puffed up. Neither have rhizines. *Hypogymnia* species are widespread, but *Menegazzia* is a speciality of Britain and Ireland's temperate rainforests, and a real gem of a find, with neat round holes in the upper surface (🔍).

Hypotrachyna (**K**, **L**) Loop Lichens' lobe edges loop around, leaving neat, dark round spaces between the lobes (arrows). Upper surfaces lack white patterning, and rhizines are branched in twos or like bottlebrushes. One small species often found on twigs is yellowish green (*H. sinuosa*, **K**) and all the others are greyish green or bluish grey (*H. afrorevoluta*, **L**).

'**Melanelia**' is an old name for a group of brownish to greenish Camouflage Lichens, which tend to blend into bark or rock surfaces. Learn to tell them apart by looking closely (🔍) at their reproductive structures, colours and textures. *Melanelixia fuliginosa*, pressed closely onto rocks, has dark, glossy lobes with isidia (**M**), and *Melanohalea laciniatula* on trees has thinner, looser lobes with isidia that flatten into miniature lobes (lobules) (**N**).

Physcia and relatives – Rosettes, Frosts and Fringes

Lobes of most species of Rosettes, Frosts and Fringes (above right) tend to be obviously smaller and narrower than those in the Shield Lichens (above left). Rosette, Frost and Fringe Lichens and their close relatives are foliose species, mostly forming small rosettes with narrow lobes and a matt surface (🔍). The commonest species are often abundant in parks and urban areas, where rosettes may grow together to form extensive patches, but look for them on other nutrient-enriched or lime-rich rocks, like concrete, bird-perch rocks or limestone.

LOOK FOR THESE KEY FEATURES
- Foliose growth form
- Usually small leafy lichens, often rosette-forming
- Green algal photobiont (scratch the surface and you should see a grass-green layer below the cortex, 🔍)
- Matt upper cortex (🔍)
- Lobe width usually about 1mm and less than 2mm
- Some turn green when wet – look for the water drop symbol in the Species Descriptions
- Sometimes with conspicuous projecting rhizines or cilia (hairs at lobe margins)

If you spot the bright yellow-orange Sunburst Lichen *Xanthoria parietina*, make sure to look for the small bright pale grey of these Rosette, Frost and Fringe lichens that often grow with it: (1) *Physcia adscendens*; (2) *P. tenella*.

Physcia (**A**, **B**) species mostly form pale grey rosettes. Look first for the presence of hair-like structures on lobe edges (cilia) and then the position of clusters of soredia to identify them. *P. adscendens* (**A**) has hair-like cilia and soredia hidden under inflated hoods on its lobe tips. *P. caesia* (**B**) has bluish-coloured mounded soredia on the surfaces of the lobes.

Physconia species are called the Frost Lichens because they have a coating of fine white crystals (**C**) like icing sugar (pruina) on their upper surfaces visible when dry (💧). They turn green when wet (**D**).

Phaeophyscia species are called Shadow Lichens, as they have muted brownish-grey colours and very flat lobes pressed closely to the surfaces they are growing on (**E**). These features, along with their small size, make them easy to overlook. They turn greener when wet (**F**). Look for projecting rhizines along their lobe edges (💧).

Anaptychia species are most common on coastal rocks, with *A. runcinata* (**G**) forming large dark brown rosettes when dry, and becoming green when wet. *A. ciliaris* (**H**) is much less common, forming small tufts. Its minutely roughened surface makes it extremely matt in appearance.

WATCH OUT for some narrow-lobed *Parmelia* relatives that do not belong to the *Physcia* group, particularly on old wood, pine bark and in the mountains: check with your hand lens – they will all have slightly shiny upper surfaces and grow in acidic situations, like *Parmeliopsis ambigua* (**I**) on old stumps or wood or *Arctoparmelia incurva* (**J**) on wood or silica-rich rock. These have narrow lobes, but are not closely related to the Rosettes, Frosts and Fringes.

A Baker's Dozen of important lichen groups

Peltigera – Pelt Lichens

There are 20 species of Pelt Lichens in Britain and Ireland, mostly large and conspicuous lichens in mossy places. Individual lobes are often 1–2cm across, with thalli up to the size of dinner plates, but often saucer sized. These lichens have cyanobacterial photobionts, and most require liquid water to grow. Like other cyanobacterial lichens, they are brown and grey, darkening when wet. A few species are symbiotic with green algae, only having their cyanobacteria in small patches on their upper surfaces or internally; these species are brilliant green when wet.

TOP TIP Pelt Lichens have **no lower cortex, but do contain a network of veins and most have conspicuous rhizines (hair-like or root-like attachment structures) on the lower surfaces** of their lobes. You will need to get in the habit of lifting a portion of the thallus up, flipping it over and gently removing attached moss to see the veins and rhizines (🔍).

LOOK FOR THESE KEY FEATURES
- **Foliose growth form**
- Large leafy lichens, often rosette-forming
- Cyanobacterial or green algal photobiont
- Lobe size usually near 1–2cm
- Lower surface cottony or felty, without lower cortex
- Lower surface with a network of pale or dark veins and rhizines (🔍)

TO IDENTIFY INDIVIDUAL SPECIES, INVESTIGATE
- Texture of upper surface: smooth or with raised bumps or fine felty hairs (🔍)?
- Rhizines: pale or dark, simple or branching (🔍)?
- Veins: pale or dark, raised or flat (🔍)?

Most Pelt Lichens have a **cyanobacterial** photobiont and are grey to brown in **colour**, like the common *Peltigera membranacea* (dry, **A**), becoming darker when wet (**B**, *P. hymenina*).

74 The pathways

A few Pelt Lichens have **green-algal** photobionts and therefore become bright green when wet. *Peltigera leucophlebia* (**C** and **D**) has ruffled lobe edges. Some green-algal species have cyanobacteria contained in external structures called **cephalodia**, which look like purplish-black spots on the surface (**D**).

Rhizines, found on the lower surface of lobes, can be long and simple like those in *P. praetextata* (**E**), or dark and fluffy as in *P. rufescens* (**F**) or anywhere in between. A rule of thumb for any large leafy lichens is to turn them upside down and have a good look at the lower surface.

On the upper surface, look for **tomentum**, a fine woolly felt mostly found near lobe tips and best seen when the thallus is dry. It can look like a dusty white covering as in *P. rufescens* (**G**) or a fine woolly covering ( ) as in *P. membranacea* (**H**).

TOP TIP Squeeze the lobe tip of a wet Pelt Lichen between thumb and finger – when you release the lichen, if there is tomentum it will show up more clearly.

The **apothecia** on Pelt Lichens are usually held erect (**I**). Only a few species have soredia or isidia. On *P. praetextata* (**J**), the tiny outgrowths on lobe edges and where the thallus is cracked or damaged are called **lobules**. These are a type of isidia, because they have a cortex continuous with the upper surface of the lichen.

A Baker's Dozen of important lichen groups

Big Browns – species of wet woodlands

If you are fortunate enough to live in or visit temperate rainforest, some characteristic groups of lichens can be conspicuous and abundant. Look out for large, loosely attached examples on mossy trees, especially willows and hazels, and you know you're in Big Brown territory. These lichens may actually be grey, black, brown or even green, but they all have cyanobacteria instead of or in addition to green algae and so they rely on lots of liquid water for photosynthesis.

These Big Browns bring together a collection of lichens that share similar habitat requirements. They are part of the so-called Lobarion, a collection of species requiring frequent wetting, year-round mild temperatures, clean air and old woodlands. If you find yourself in a moist ravine or old woodland, especially the further west you go, consider the possibility of lichens with romantic names like Moon, Felt, Kidney and Lungwort lichens along with the wonderfully named Specklebellies. The Big Browns often share habitats with some of the Pelt Lichens (p. 74) and Jelly Lichens (p. 78), so look out for those, too. The name Lobarion comes from the Lungwort *Lobaria pulmonaria*, which can be the easiest to spot, as it is lettuce-sized and bright green when wet.

TOP TIP The cardinal rule for learning to recognise anything big and leafy in wet woodlands is to check the lower surfaces carefully ()!

LOOK FOR THESE KEY FEATURES
- Foliose growth form
- Large, loosely attached lichens
- Temperate rainforests or mossy, damp woodlands
- Lower surfaces will help you identify these
- Many brown, black or dark grey colours, darker when wet
- Some greenish brown, bright grass green when wet

Sticta—Moon Lichens (**A** and **B**) have a fine and short-velvety coating of creamy brown hair on their lower surfaces (called a **tomentum**), punctuated with neat round pores to increase airflow to the interior (**B**). These pores are called **cyphellae** (arrows), defined by being lined with a thin membrane which is visible at the edges of the pore ().

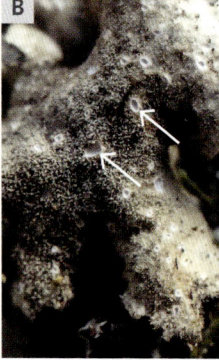

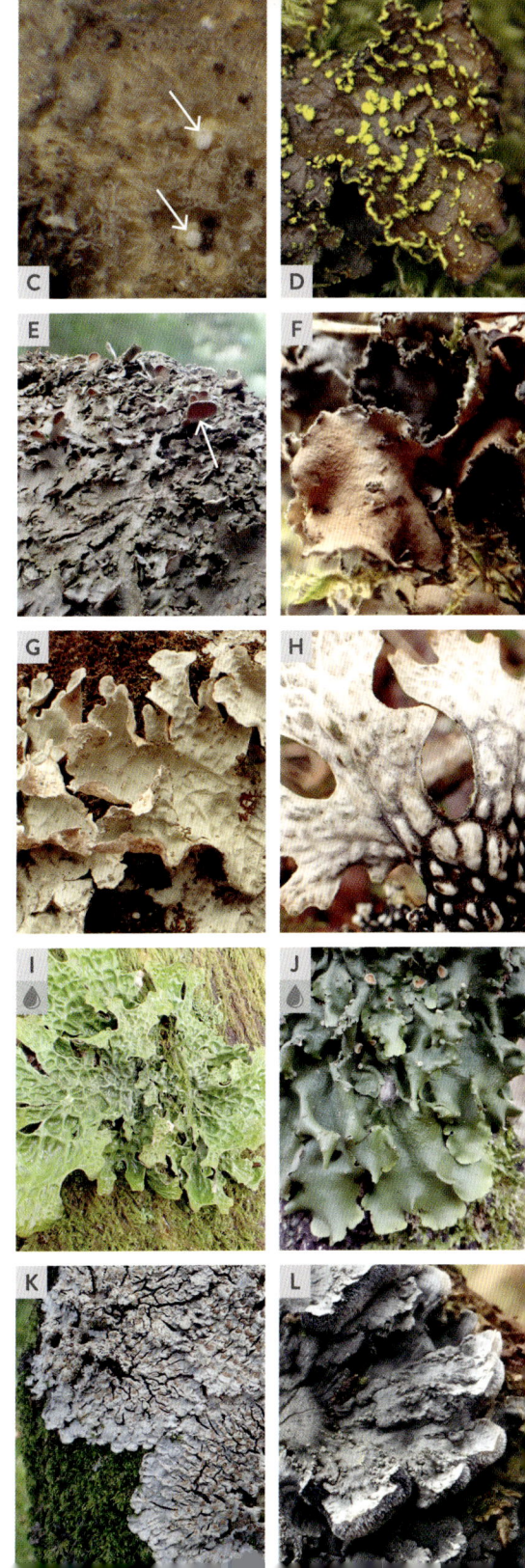

Pseudocyphellaria—Specklebellies (**C**) have tomentum and pores on the lower surfaces like *Sticta*, except the pores on Specklebellies are more like little pale patches and don't have neat bordered edges – so 'pseudo' 'cyphellae' (arrows). If you ever see brilliant yellow soredia on a large leafy lichen, this will be a *Pseudocyphellaria* (**D**). All four British species in this genus have powdery soredia, and two have yellow soredia.

Nephroma—Kidney Lichens (**E**) are not so big, but they will be found with the Big Browns. On the lower surfaces of their brown lobes, they have reddish-brown apothecia that are somewhat kidney shaped (arrow). They are smooth on their lower surfaces, without rhizines or tomentum (**F**).

Lung Lichens (**G** and **H**) have a network of ridges on the upper surface and a **tomentum** of fine short hairs forming a net-like pattern on the lower surface (**H**), darker in the centre of the thallus, mirroring the pattern of ridges on the upper surface. *Lobarina scrobiculata* (**G**) has rounded lobes; *Lobaria pulmonaria* (**H**, showing undersurface) has wide gaps between lobes.

Pale brownish to bright green when wet. *Lobaria pulmonaria* (**I**) and *Ricasolia virens* (**J**) can be greenish brown, but both turn bright green when wet. The green colour tells you they have green algal photobionts, but they also have internal pockets of cyanobacteria for nitrogen fixation – clever!

Pectenia—Felt Lichens (**K** and **L**) are neither big nor brown, but look for them along with Big Browns. They form neat rosettes and sit on a thick, richly branched weft of pale to dark grey or black fungal hyphae (), like a sponge (**L**). This is the so-called **hypothallus** (hypo = below).

Jelly Lichens

Several unrelated groups are referred to as **Jelly Lichens**, as their internal structure is made of a gelatinous matrix. These lichens do not have an internally layered structure. The jelly matrix is produced by their filamentous cyanobacterial photobionts, and there are fungal threads running through the jelly but little other internal structure. Some species have a single cell layer for a cortex, making their upper surfaces slightly smoother than species which have no cortex at all.

Most Jelly Lichens are dark green, brown or blackish, and many of them change considerably when wet, becoming rubbery as the jelly-like matrix fills with water and swells. You'll need to observe them when dry to identify them, as some diagnostic features are obscured when swollen with water. In regions of temperate rainforest there are many different species; these are nitrogen-fixing lichens.

TOP TIP Jelly Lichens have cyanobacterial photobionts and require liquid water to become photosynthetically active, so **look for these lichens in mossy, damp places, on shaded tree bases or horizontal boughs, or on moist soil or walls**.

The largest group of jellies is the *Collema* family, mostly comprising foliose lichens, but there are other unrelated Jelly Lichens including the coastal fruticose lichen *Lichina*. The classification of the *Collema* family has changed recently, so you'll see new genus names being used for some of their members.

See *Lichina*, *Enchylium*, *Lathagrium* and *Scytinium* on pp. 96 and 142–44.

LOOK FOR THESE KEY FEATURES

- **Fruticose** or **foliose growth form**
- Dark brown, black or dark green in colour
- No internal layering found in cross-section
- Habitats where liquid water is available
- *TO IDENTIFY INDIVIDUAL SPECIES, INVESTIGATE*
- Wet versus dry textures: how much swelling?
- Surface features when dry: wrinkled or smooth, with isidia (🔍)?
- Felty pale brown lower surfaces in some western species (🔍)?
- Shapes of isidia: spherical, ear-like, spine-like (🔍)?

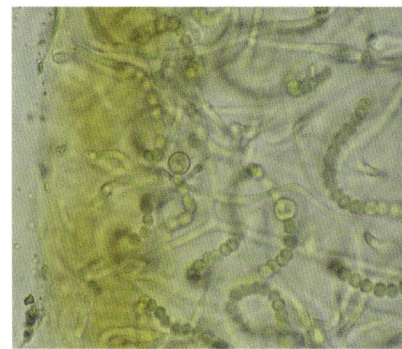

Lathagrium auriforme, 400×, mounted in water under the microscope. The chains of greenish bead-like cells are the cyanobacterium photobiont *Nostoc*, with larger cells where nitrogen fixation takes place. The straighter clear filaments are the fungal threads (hyphae), all embedded in a polysaccharide jelly. The upper surface (left of image) is pigmented, but without different anatomy.

Lepra, *Pertusaria*, *Ochrolechia* and relatives – Supersized Crusts

Considering these crustose species as a single group is sensible, because together they have some useful features to recognise before you begin to tackle them individually, and they are close relatives. Most of these species are thick and chunky crustose lichens, with thalli that are large, sometimes growing over mosses. Sticking to the 'large' theme, soredia are individually large and granular, easily visible individually with a hand lens, and soralia (distinct clusters of soredia) are correspondingly prominent. A few species also have isidia, also typically of a large size. When present, apothecia are often prominent and large.

The outer growing edge of these lichens is frequently zoned, or made of different coloured or textured concentric rings, visible without a hand lens. The outermost zone is often white, of radiating fungal hyphae only (🔍).

TOP TIP When you see a thick and chunky crust, this group is a good place to start for an identification. However, to identify them to species level, chemical tests are often needed for confirmation. We present a few species in *Lepra*, *Ochrolechia* and *Pertusaria* that are straightforward to recognise just with a hand lens and a willingness to dabble in lichen tasting!

LOOK FOR THESE KEY FEATURES

- **Crustose growth form**
- Large lichens with green algae
- Thick, chunky thallus often with a zoned edge
- Large reproductive structures
- Large apothecia with thick margins *or* coarse soredia in large clumps (soralia) *or* thick isidia

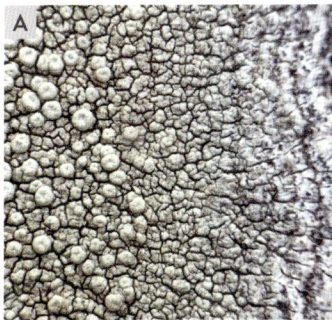

Examples of Supersized Crusts: (**A**) zoned thallus margin in *Ochrolechia parella*; (**B**) *Lepra albescens* overgrowing moss; (**C**) large apothecia with thick margins in *Ochrolechia tartarea*; (**D**) large clusters of large soredia in *Lepra amara*.

The 'Caloplaca' group – Firedots

Firedots form the largest group of orangish to yellowish crustose lichens, formerly all belonging to the huge genus *Caloplaca*. They share the same orange pigments as and are very closely related to the Sunburst Lichens (pp. 139–41). Recently, Firedots have been subdivided into different genera; nevertheless, it is a reasonable approach when you are starting out to think of all these orange crusts as 'Caloplaca' group lichens or just plain Firedots.

Firedots almost always prefer some sort of nutrient rich substrate: they can be conspicuous on urban walls, coastal rocks, monuments and on lime-rich rocks like limestone, forming a collection of orange apothecia, yellowish-orange to orange crusts, sometimes with lobed thallus margins (Lobed Crustose), or even just yellowish or orangish dusty smears.

LOOK FOR THESE KEY FEATURES
- **Crustose growth form**
- Many exhibit orange colours
- Some have lobed edges (🔍, see below)
- Apothecia mostly orange, tending to be neat and mostly flat (🔍)

A number of Firedots, such as *Variospora flavescens* (**A**), have **lobed thallus margins**. These lobes are still completely attached to the substrate and cannot be peeled off with a knife or fingernail, so they remain true crustose lichens. Look for the shape of lobes, presence of powdery pruina (like icing sugar dust) and mode of reproduction (🔍); *Calogaya decipiens* (**B**) also has lobed thallus margins along with coarse yellow soredia on inner lobe tips (🔍).

80 The pathways

Some Firedots are only spotted because of their **orange apothecia** contrasting against a dull thallus, like *Blastenia crenularia* (**C**) on silica-rich rocks or *Athallia holocarpa* (**D**) on lime-rich rocks.

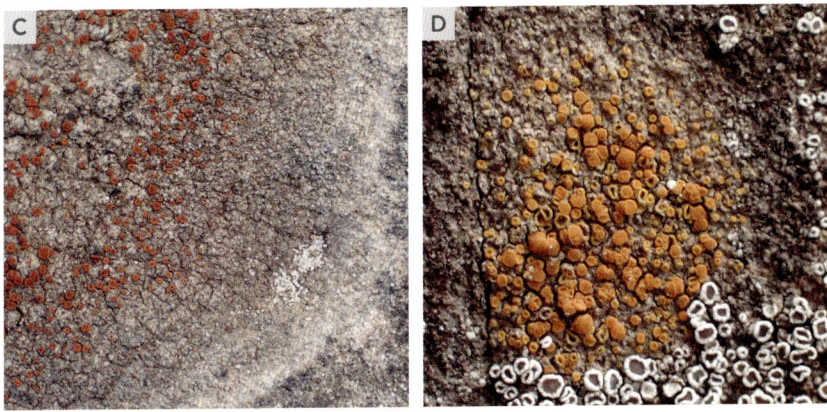

A handful of Firedots appear to be made up almost entirely of **soredia**, including the abundant – even weedy – *Flavoplaca citrina* (**E**), found in cities and towns, on stone and on brick walls. It can sometimes form extensive colonies which can look like a wash of yellow paint. Up close, the soredia are irregular and lumpy and you might find orange apothecia (**F**) (🔍).

> **WATCH OUT** for mustard-yellow thalli made of convex, smooth granules (🔍), which belong in a different group altogether, *Candelariella*. Both '*Caloplaca*' and *Candelariella* (p. 193) include bright yellow to orangish crusts, but with practice it's not difficult to tell apart a few of the common species. The most reliable method to distinguish them is by using a simple chemical test (see **Resources and Next Steps** and **Further Reading**, p. 239), which you may want to investigate later in your lichen journey, so you can be confident about your identifications.

A Baker's Dozen of important lichen groups

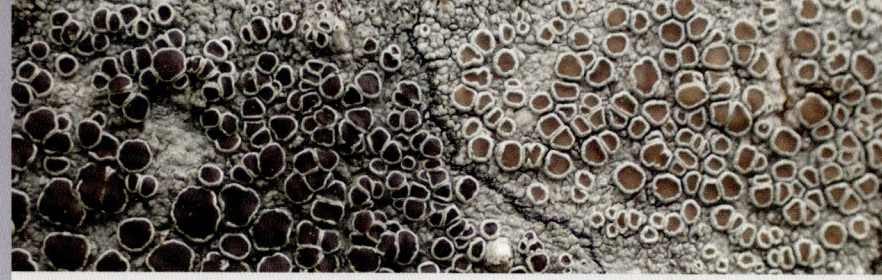

The 'Lecanora' group – Rim Lichens

The **Rim Lichens** are a large group of crustose lichens with algae in the margins of their apothecia. The presence of algae here (arrow) means that the margins look very similar to the thallus, and so they are sometimes called **thalline margins**. You might think of them as the 'jam tart' lichens. Most of the species with thalline margins on their apothecia were once included in the genus *Lecanora* and some still are, so using that name will still get you to the right place with a bit of sleuthing. Other features of the genus are microscopic and chemical and beyond the scope of this book, so we present a few common species to get a sense of the group as a whole.

Rim Lichens can be encountered on a variety of substrates, from trees and worked timber to rocks, mortar and concrete. For those just starting out, these lichens can be identified at this group level, though a few common species are introduced in this book.

LOOK FOR THESE KEY FEATURES

- **Crustose growth form**
- Apothecia with margins (rims) that match the colour of thallus and often contrast with the disc

Some very common species appear to have no thallus at all, like *Myriolecis dispersa* (**A**), and many, such as *Lecanora polytropa* (**B**), are a subtle but telling colour of yellowish green due to the presence of usnic acid, the same chemical that makes *Usnea* species have that same colour.

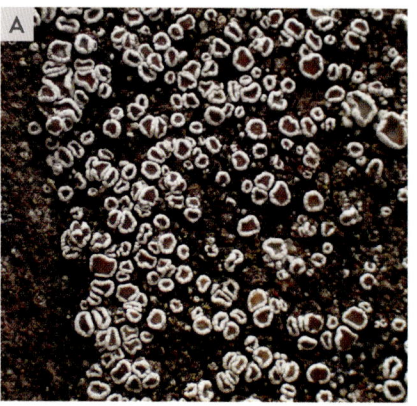

Lecidea and other black dots on rocks – Wine-gum Lichens

The Wine-gum Lichens are the counterpoint to the Rim Lichens. They are known as Wine-gums because their apothecia are without thalline margins. Instead, the margins are typically the same colour as the disc – a crustose lichen with tiny wine-gums scattered on it. Historically, Wine-gum Lichens were all considered part of one huge genus, *Lecidea*.

Like the Rim Lichens in *Lecanora*, the Wine-gum Lichens that were historically included in *Lecidea* have been split up into many more genera. When you are getting going with lichens, it's a great habit to begin to observe the margins of apothecia and to recognise these two groups as a starting place for identification. The margins in Wine-gum Lichens are described as **lecideine** (meaning 'like *Lecidea*') or as having **non-thalline** margins.

These lichens can be common, conspicuous and abundant, but microscopy and chemistry are required to confirm the identifications of most of them. We include a handful of individual species in this book that can be straightforward to identify, including *Lecidella elaeochroma* (p. 210), *Mycoblastus sanguinarius* (p. 211) and *Rhizocarpon reductum* (p. 212).

> **LOOK FOR THESE KEY FEATURES**
> - Crustose growth form
> - Often black apothecia with black margins

This page shows a few common groups of Wine-gum Lichens, particularly those of silica-rich rocks in the uplands. See also a set of species that grow on iron-rich rocks, whose thalli often are red or orange colours (see p. 233).

Porpidia and *Lecidea* (**A**) can be difficult to tell apart without microscopy, and they often grow together.

Rhizocarpon (**B**) often has thalli with the surface broken into minute and neat islands (areoles).

Fuscidea (**C**) have only brown pigments in their thalli and apothecia. *F. cyathoides* can be very abundant on upland rocks, forming mosaics. You may be able to spot the slightly paler margins of its apothecia with your hand lens.

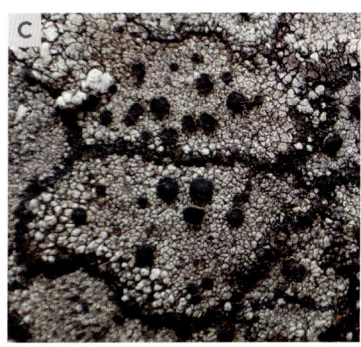

A Baker's Dozen of important lichen groups

Graphids and *Arthonia* and relatives – Scripts and Commas

The Script and Comma Lichens belong to two groups of unrelated fungi, the Graphid group (a single family) and *Arthonia* and relatives (a few related families). These lichens are considered as distantly related to each other as birds are to mammals; they are not only in different families, but also in different orders within different classes of life. Nevertheless, some of their members produce similar and distinctive elongate apothecia, like tiny scribbles on bark. These long, narrow apothecia are called **lirellae**. For the sake of preliminary identifications, we can include them here in the same group.

Plenty of non-lichenised fungi make lirellae, too, so you should first check that you are definitely looking at a lichen. How? Both Script and Comma lichens have a green algal photobiont called *Trentepohlia*, which has distinctive bright orange pigments, so instead of a green layer below the upper surface, it is orange (see image below).

TOP TIP   If you scratch the thallus with your fingernail, you will see a yellowish-orange smudge.

In Britain and Ireland, the Graphids tend to be more conspicuous, often with pale thalli contrasting in colour to their black lirellae. Once you start looking closer with your hand lens, you may find more Scripts and Commas. They can can be quite diverse the farther west you are (towards the temperate rainforest), particularly on smooth bark.

See *Arthonia* and *Graphis* (pp. 223–25).

LOOK FOR THESE KEY FEATURES
- **Crustose growth form**
- Apothecia shaped like lines, irregular squiggles or star-shapes, called **lirellae** (🔍)
- Yellowish-orange scratch (🔍)
- Often on smooth bark
- Some have black lip-like margins to their lirellae (🔍)

An *Arthonia* group species (*Felipes leucopellaeus*) with short and irregular lirellae and a scratch showing the orange-pigmented photobiont layer just under the upper surface.

LICHENS BY FORM AND COLOUR

In this section of the book, you will find lichens organised by form and colour. The lichens are first divided into the main growth forms – fruticose, foliose and crustose – and within those, they are organised by colour and other shared similar features. The Species Descriptions are arranged following this organisation, so you should find species that look similar and share similar features together when you flip through the pages.

We have created three Quick Guides with thumbnail images of the lichens, one for each growth form – fruticose, foliose and crustose. *Follow the steps on the next page to help you narrow down your options when seeking an identification.*

When you approach a lichen-rich habitat like this tree trunk, start by sorting lichens into growth forms, and then sort by colour, branching pattern and reproductive structures. Before you know it, you'll have made a great start at identifying the lichens you can see.

STEP 1 GROWTH FORM
Check the flap on the front cover for a reminder of the growth forms, and then go to the relevant Quick Guide at the start of each growth form section. Use the icons in the page margins to find the relevant section.

STEP 2 COLOUR
Once you have decided on your growth form, check the relevant colour sections in the Quick Guide to find sets of lichens with similar colours. We separate out obvious colours including bright green, orange and yellow or black and dark brown; the muted pale, greyish and greenish colours are grouped together.

Please note that although colours can be helpful, lichens are greener or darker when wet, and they can be variable, sometimes depending on where a particular thallus is growing. There is no perfect way to sort lichens into easily recognisable sets, but colours can be one of the most immediate things we notice. We do offer similar species in almost every Species Description, so it's always worth checking the closest match or matches in the Quick Guides for other possibilities.

STEP 3 SIZE AND FEATURES
Look for other features that your lichen may have, such as its pattern of branching, shape or size of lobes, or reproductive structures. Be curious, use your hand lens; take a small piece in your hand and turn it over to examine lower surfaces if there are any. Don't forget, if your specimen has no reproductive structures, it may not be mature and might be impossible to identify, so spend a moment trying to locate these. Use the flap on the back cover for reminders of different reproductive structures.

STEP 4 SPECIES DESCRIPTIONS
When you think you have found a match for your lichen, check its Species Description page for further information which may help you confirm your identification. If you find conflicting information – for instance, if it says that the species only grows on the tops of mountains and you are on the coast – look back again to the Quick Guide and check for similar-looking species in the other colour sections or leaf through the Species Descriptions for other possibilities.

FRUTICOSE LICHENS

We separate lichens that have this growth form by a combination of colours and branch shapes as shown on the Quick Guide of thumbnail images on the next page. First separate out lichens with bold colours, and for more muted colours, check branch shapes. When you find a best match using the thumbnails, go to the page number for the **Species Description**. Be sure to check the Similar Species headings for other species that may be presented there or elsewhere in the book.

1. Bold colours
- There are only a small number of fully **orange** fruticose lichens, but sometimes species that are more often greyish or greenish can be orange. We have put these in the orange section, but we also reference them in the other sections according to branch shapes.
- Fruticose lichens that are clearly **brown** to **black** are also usually easily separated by colour, but may be hard to spot! In heathlands, it is easy to miss the darker species of *Cetraria*, *Alectoria* and *Cladonia*, so sit awhile, have a cup of tea, and check the ground!
- Most fruticose lichens you encounter will be subtle shades of greyish, to pale greens, yellows or even white. For these colours, check the branches.

2. Branching
- Hair-like Hair-like fruticose lichens have thin branches, usually mostly round in cross-section. *Usnea* has a central stretchy cord, and *Bryoria* and *Alectoria* do not. *Alectoria* has tiny white markings (pseudocyphellae), while *Bryoria* does not.

TOP TIP Think your lichen might be an *Usnea*? Hold two ends of a medium-sized branch and gently pull apart to reveal the inner stretchy cord.

- Many fruticose lichens are tufted. Check to see if branches are:
  - hollow by taking a piece of a branch and tearing it in two (*Cladonia*, *Thamnolia*)
  - solid (*Stereocaulon*, *Sphaerophorus*)
  - flattened (*Ramalina*, *Flavocetraria*, *Cetraria islandica*).

WATCH OUT for foliose lichens (*Evernia*, *Pseudevernia*) that grow rather bushy, and crustose lichens with stalks (*Baeomyces*, *Dibaeis*).

Fruticose lichens

QUICK GUIDE

FRUTICOSE LICHENS

BROWN TO BLACK

Lichina (p. 96)

Cetraria (p. 92)

Cetraria (p. 93)

Cornicularia (p. 95)

Bryoria (p. 97)

Cladonia (p. 94)

YELLOW TO ORANGE

Teloschistes (p. 90)

Usnea rubicunda (p. 91)

Sphaerophorus (p. 118)

PALE WHITISH OR GREY TO GREENISH

BRANCHES HAIR-LIKE

Usnea (pp. 98–100)

Bryoria (p. 97)

Alectoria (p. 101)

PALE WHITISH OR GREY TO GREENISH

BRANCHES TUFTED OR MAT-FORMING, *NOT* HAIR-LIKE

— *Branches hollow*

Pixie Cup *Cladonia* (pp. 102–105)

Pointed or simple *Cladonia* (pp. 106–11)

Reindeer lichen *Cladonia* (pp. 112–14)

Thamnolia (p. 115)

— *Branches solid*

Sphaerophorus (p. 118)

Stereocaulon (p. 116)

Bunodophoron (p. 119)

Rocella (p. 120)

— *Branches flattened or strap-shaped*

Ramalina (pp. 121–23)

Flavocetraria (p. 124)

Rocella (p. 120)

Cetraria (p. 93)

Upper and lower sides different colours (foliose)

Evernia (p. 130)

Pseudevernia (p. 131)

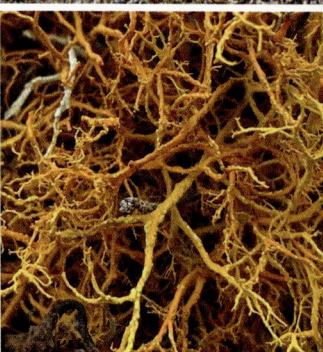

Teloschistes flavicans Golden Hair Lichen

HOW TO SPOT Look for bright orange bushy lichens – especially in high light situations, e.g. on wayside hawthorns or coastal rocks.

DESCRIPTION Fruticose. Thallus unmistakable, yellow to orange bushy lichen up to 3–4cm high, tufted or slightly drooping with fine irregular branches and spine- or claw-like tips. *Reproduction:* **Powdery soredia** are borne in dot-like clusters (soralia) (🥄).

WHERE On sunny, windy rocks on offshore islands and on twigs of wayside trees in the south.

NOTES This species is rare and should not be collected; photos are sufficient for confirmation. As with all orange lichens, the more sunshine, the darker orange the colours, so this species may be greyish in shadier sites.

SIMILAR SPECIES

T. chrysophthalmos Golden Eye Lichen (**A**) is the only other bright yellow to orange bushy lichen in Britain and Ireland, but differs in having dark orange sunflower-like apothecia, rimmed in tiny radiating branches like rays of sunshine. Both *Teloschistes* species are apparently on the increase as temperatures rise.
(**B**) ***Sphaerophorus globosus*** (p. 118) can be orangish, but usually has grey branch tips.

Usnea rubicunda Red Beard Lichen

HOW TO SPOT Tufted or long-drooping and stringy, often yellowish green with orange to red parts at base, rarely completely orange or reddish.

DESCRIPTION Fruticose. Thallus yellowish green to subtly reddish orange, especially with orange tones on largest branches, forming tufts to 5cm or long-drooping up to 10–20cm. Branches coarse and wiry, with abundant white warts ( ). *Reproduction:* Isidia produced from tiny wart-like bumps.

WHERE On trunks and branches in the west.

NOTES Different *Usnea* species often grow together, so look for the orange tones that stand out against the usual yellowish-greens. This species is very variable in colour, from mostly yellowish green to rarely completely bright orange-red. Even in mostly yellowish-green examples, there will be some orangish colour mostly on larger branches at base, but some specimens will be vexing due to only subtle orangish tones.

Cetraria aculeata Spiny Heath Lichen

HOW TO SPOT Hard to spot! The sort of lichen you miss because of its dark colour. Dark, densely branched spiky-looking patches on soil in heaths.

DESCRIPTION Fruticose. Thallus reddish brown to brown-black at visible tips, paler and greenish within the cushions or in deep shade, spreading up to 15cm, but usually smaller, up to about 2–4cm tall. Branches fine, 1–2mm wide, glossy, hollow, usually slightly flattened in cross-section and richly and repeatedly branched, with spiky tips; tiny white pits or holes in the thallus (sunken pseudocyphellae,), which are small on upper branches and deeper and larger below on older parts. *Reproduction:* **Apothecia rare**, smooth to irregular discs the same colour as the branches, located on tips of branches.

WHERE On peaty soil and rarely tree bases or stumps in heath, but also spoil piles, dunes, shingle and mossy screes.

SIMILAR SPECIES

Although the texture and branching is similar with *Cetraria islandica* (opposite), in reality, you won't confuse them. *Cetraria aculeata* forms low, dense, spiky mats and branches are much finer, round in cross-section and hollow. Some *Cladonia* species (see p. 94 and *C. furcata*, p. 110) can be brown in full sun, but they are not glossy textured. *Pseudephebe pubescens* (p. 95) has finer, kinked, hair-like, greenish-brown, solid branches and grows on the tops of boulders in the mountains, looking a bit like matted hair.

Cetraria islandica Iceland Moss

HOW TO SPOT Often tricky to spot, as its dark tones tend to camouflage it among the shadows of vegetation, in which it is often loosely tangled. Break for your lunch in a patch of short heath and you may just find you've sat on it.

DESCRIPTION **Fruticose.** Thallus colour ranges from reddish to greenish brown to almost black, greener when wet, with a distinct glossiness. Upright, strap-shaped and inwardly rolled lobes up to 1cm wide (usually 2–5mm), often up to 5–6cm tall/long with a few side branches and tiny spiky side projections (). One side has scattered paler spots (pseudocyphellae), and the brownish pigment is much reduced or absent in shaded parts of the thallus tucked among plants, where it is pale beige to greenish. *Reproduction:* **Apothecia rarely seen**, but microscopic clonal spores can be produced within the tiny projections (pycnidia).

WHERE On soil in hilly and mountainous areas. Common in lichen-rich heaths in the Scottish Highlands, but less so the further south you go.

NOTES This species has long been used as a remedy for sore throats and is still sold in bulk in some countries for this use. It also forms an important food source for reindeer/caribou.

Cladonia cervicornis Browned Pixie Cup

HOW TO SPOT Dense cushions of tiny squamules with short wide cups – sometimes with newer cups stacked inside older ones.

DESCRIPTION Fruticose. Pale to brownish green to brown, dense mats of indented squamules 2–3mm long with neat greenish-brown to brown wide-flared, goblet-shaped cups. Sometimes tiered goblet-shaped cups have new cups arising from the centres of older ones. Stalks of cups smooth. *Reproduction:* **Apothecia**. Older cups may have small brown apothecia arising at edges.

WHERE On thin or mossy soils in dunes and heaths.

Cladonia gracilis Smooth Horn Lichen

HOW TO SPOT Slender smooth and graceful tufts of unbranched stalks tipped with delicate goblets are often very camouflaged within vegetation and among other lichens; usually only becomes visible when you have your nose to the ground.

DESCRIPTION Fruticose. Thallus brown (in full sun) to greenish (in shade), up to 6cm tall, of rather fine and graceful smooth stalks. Often in dense clumps up to 5cm wide or mixed with other *Cladonia* species. *Reproduction:* **Apothecia, brown**. Tips of stalks are occasionally narrowly expanded into delicate goblets, sometimes dotted with apothecia, and sometimes with new branches growing from edges of goblets.

WHERE On acid soil in heaths and uplands.

SIMILAR SPECIES

Look out for little-branched forms of *C. furcata* (p. 110), which can be brown or green, but usually has some forked tips. *C. cervicornis* (above) has shorter, wider cups and conspicuous mats of scale-like leafy lobes (squamules).

BROWN TO BLACK FRUTICOSE

Cornicularia normoerica Bootstrap Lichen

HOW TO SPOT Look for short spiky blackish tufts on exposed boulders in wild and windy uplands.

DESCRIPTION Fruticose. Thallus brown-black, forming patches a few centimetres in extent, with shiny smooth flattened branches up to 2cm high and up to 1mm wide, which often divide into two equal parts, narrowing at the tips. *Reproduction:* **Apothecia** of shiny discs up to 6mm wide, found at the end of the branches; can be absent or abundant.

WHERE On silica-rich coarse-grained rocks, found mostly in Scotland, northwest Ireland and northwest Wales in upland or mountainous areas, well exposed to sun and wind.

NOTES There is only one species of *Cornicularia* in Britain and Ireland. Often found with Yellow Map Lichen *Rhizocarpon geographicum* (p. 192) and Rock Tripe Lichens *Umbilicaria* species (pp. 155–56).

SIMILAR SPECIES

Pseudephebe pubescens has finer, more richly branched thalli and tends to creep in low blackish mats across rocks in similar habitats.

Fruticose lichens 95

Lichina pygmaea Black Seaweed Lichen

HOW TO SPOT Look for wide-spreading, turf-like patches of matt black growing in the mid-shore, often among barnacles on coastal rocks.

DESCRIPTION Fruticose. Thallus blackish, up to 1cm tall, forming wide-spreading patches 10–20cm across; branches flattened, dividing in 2s. ***Reproduction:* Apothecia** (arrows) are tiny and globe-shaped, up to 2mm, opening by a pore at the tips.

WHERE On exposed sunny and sheltered seaside rocks, or tucked into crevices where wave action is higher. Often found among barnacles, the black Tar Lichen *Hydropunctaria maura* (p. 232) and the orange *Flavoplaca marina* (p. 190).

NOTES The blackish colour hints that this lichen contains cyanobacterial photobionts. In fact there are two different cyanobacteria present. Each is optimised for photosynthesis at different stages of the tide – one while the lichen is submerged at high tide, and the other when exposed at low tide.

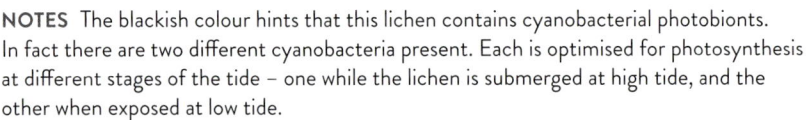

SIMILAR SPECIES

L. confinis is similar, but – paradoxically – smaller, up to 0.5cm tall with tiny lobes round in cross-section and often forming less conspicuous patches. The peculiar name of *L. pygmaea*, suggesting small size, is relative to the seaweeds with which it is often found, and to which it was first thought to be related.

HAIR-LIKE FRUTICOSE

Bryoria fuscescens Horsehair Lichen

HOW TO SPOT Brown and hair-like in soft beard-like tufts on conifers and birches, fenceposts or rocks.

DESCRIPTION Fruticose. Pale to dark or greyish-brown, very fine, smooth hair-like strands in drooping tufts 5–20cm long, with wide-angled branching. *Reproduction:* **Powdery soredia** in dense patches (soralia) arising in fissures and expanding to become wider than the branches.

WHERE On trunks, branches and twigs of birch and conifer trees, fenceposts and silica-rich rocks in clean-air areas.

NOTES Prefers acid conditions and is very sensitive to ammonia and other air pollution. Now in decline in Britain and Ireland due to excess atmospheric nitrogen. Hair lichens are a key source of winter food for reindeer and caribou in forests of northern countries when the ground is snow-covered.

SIMILAR SPECIES

Other dark *Bryoria* species are uncommon; most have many short, spiky side branches and none have pseudocyphellae. *B. bicolor* (not shown) is erect on mossy trunks and rocks, with two-toned branches black at the top, greenish near the base. *Alectoria nigricans* (right) in high-elevation heath is up to 5cm in height, with tiny pale pseudocyphellae (⊙).

Fruticose lichens 97

HAIR-LIKE FRUTICOSE

young soralia

mature soralia

Usnea subfloridana Boreal Beard Lichen

HOW TO SPOT Yellowish-green tufts on twigs and branches with plenty of light and airflow.

DESCRIPTION Fruticose. Thallus yellowish green but black near the holdfast, bushy or slightly drooping, but not more than 1.5 times longer than wide, up to 10cm long; branches cylindrical throughout. *Reproduction:* Isidia are produced in round, convex clusters; these remain visible even on old branches.

WHERE On twigs and branches with plenty of light and airflow, especially in the north and west.

NOTES *Usnea* species are very sensitive to all kinds of air pollution and are only found in clean-air areas. These lichens may grow quickly, and even though they may be relatively large, you will need to check the base and find mature soralia to identify most species.

SIMILAR SPECIES

U. wasmuthii is very similar and often occurs with *U. subfloridana*. It forms oval patches of soredia (soralia) without isidia on mature branches (). Young branches (**A**) start out with small patches of isidia, but soon shed these and soralia elongate to ovals along the branches (**B**). The black base has very fine brickwork-like cracking, but this can be tricky to see (; p. 69, E).

A

B

98   The pathways

base

Usnea cornuta Inflated Beard Lichen

HOW TO SPOT Short yellowish-green tufts on bark in the south and west, spiky-looking up close.

DESCRIPTION Fruticose. Thallus yellowish green to grey green, bushy, usually up to 6cm, with branches mostly at 90-degree angles, and so appearing spiky; larger branches constricted where they meet the main stem (, arrows). Basal part near holdfast pale (), to orangish or brownish. *Reproduction:* Isidia are abundant in small clusters (soralia) of differing sizes, which are densely packed on the smaller branches.

WHERE On trunks of trees; also on wooden railings, branches and twigs. Probably the most abundant *Usnea* in the south and the west.

NOTES When dry, the main branches will crack and collapse with pressure from a thumbnail due to their loosely woven layer below the outer cortex.

SIMILAR SPECIES

U. cornuta is one of several species with inflated branches because of having a loosely woven medulla (cottony-white inner layers of the lichen), and by far the commonest of those. Look carefully for isidia on young branches; some other rarer tufted species only have clusters of soredia (soralia) and never isidia, such as Witch's Fingers *U. esperantiana*, which has curled, gnarled tips and appears to be spreading in the south.

pseudocyphellae

Usnea articulata String of Sausage Lichen

HOW TO SPOT Forming tangled mats in trees and shrubs, with large, inflated sections on mature branches, constricted like sausages on a string – unmistakable.

DESCRIPTION Fruticose. Thallus yellowish green, forming drooping tangles up to 1m long but often smaller. Branches up to 5mm wide, with constrictions where main branches meet; the central cord is sometimes visible between inflated sections, inviting its common name. Look closely and you may spot faint pale dot-like or comma-shaped pseudocyphellae (; arrows on inset photo). *Reproduction:* **Isidia** produced from tiny soralia.

WHERE On twigs and branches in tops of trees and shrubs in well-lit sites mostly in the southwest of England and southern Ireland and Wales, and rarely scrambling on the soil in dunes in Norfolk.

SIMILAR SPECIES

U. rubicunda (**A**, p. 91) and (**B**) *U. ceratina* (Warty Beard) can also be long-drooping, but these have many fine side branches at right angles to main stems. *U. ceratina* has a wiry look and a pinkish layer between the outer cortex and central cord (break a branch to see inside) with whitish warts on the branches ( ).

Usnea dasopoga Fishbone Beard Lichen

HOW TO SPOT Look for long-drooping, yellowish-green tufts on conifers and birches.

DESCRIPTION Fruticose. Thallus yellowish green, up to 15cm long, almost always more than 3× longer than wide, of cylindrical branches with a fishbone pattern to the branching, with long straight main branches (the spine of the imaginary fish) and many smaller perpendicular branches (the ribs). Black holdfast. *Reproduction:* Fragmentation of fine side branches and isidia.

WHERE On acid bark, trunks, branches and twigs, especially in clean-air areas in north.

Alectoria sarmentosa Common Witch's Hair

HOW TO SPOT Look for long-drooping strands of yellowish green on conifers, cliff faces, or 'silly string' lying flat on mountain soil.

DESCRIPTION Fruticose. Thallus yellowish-green, up to 30cm or more long, with branches irregular in cross-section and branching at wide angles; lacks the strong internal cord of *Usnea*. Look for small raised white patches (, pseudocyphellae, arrows) on the branches. *Reproduction:* Presumably mainly by fragmentation.

WHERE On bark or rock in acid situations. Especially on pine bark, trunks, branches and twigs, and rock faces especially in clean-air areas in the north. The form that grows on soil is found on high mountain heaths, where the vegetation is clipped low by the wind. Rare, but exciting to find!

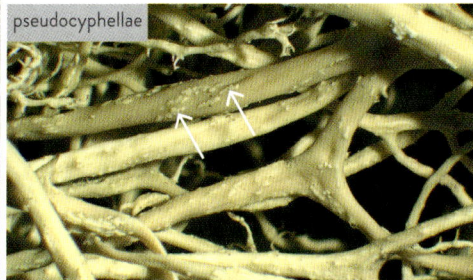
pseudocyphellae

SIMILAR SPECIES

From a distance *U. dasopoga* looks very similar, but its fishbone branching is distinctive. Pale specimens of *Bryoria* lack the tiny raised white pseudocyphellae of *Alectoria*.

Fruticose lichens 101

Cladonia chlorophaea Mealy Pixie Cup

HOW TO SPOT Usually small collections of green-grey or brown-grey stalks with wide goblet-shaped tops, often on tree bases, stone walls or thin well-drained soil.

DESCRIPTION Fruticose. Greenish to brownish-grey, goblet-shaped stalks to about 15mm, arising from mats of small scale-like squamules which are white on the lower surfaces and indented on their tips. Stalks have a tiny ring of smooth cortex 1–2mm high at the base, and flare to form wide cups. *Reproduction:* **Soredia or granules** (larger soredia) sometimes with a thin cortex, so appearing more compact or even smooth () on upper parts of stalks and inside cups. These can be individually seen with a hand lens. Brown apothecia arise from the edges of cups, but are not common.

WHERE On tree bases, mossy walls and monuments, and well-drained acid soils.

NOTES If overgrown by other lichens or vegetation, new growth may be initiated from the edges of the cups, creating tiered cups.

SIMILAR SPECIES

C. fimbriata has more suddenly flared cups often with tiny teeth at the edges, and fine floury soredia (impossible to see individually with a hand lens) along the entire length of the stalks, without a ring of cortex at the base. *C. pyxidata* and *C. pocillum* (p. 103) have flattened granules inside the cups and are found on less acid substrates. *C. cervicornis* (p. 94) has no soredia or granules on the stalks or within the cups.

Cladonia pocillum Rosette Pixie Cup

HOW TO SPOT Neat, well-formed rosettes of leafy scales with short-stalked, wide-flaring cups on lime-rich soil, on moss growing on limestone, dunes or old mortar.

DESCRIPTION Fruticose. Green-grey, dense and usually neat mats of overlapping, horizontally spreading squamules, forming radiating rosettes, topped by short, goblet-shaped stalks to about 15mm. *Reproduction:* **Coarse granules** (larger soredia) with a thin cortex, so appearing compact or even smooth () on upper parts of stalks and cups; these become flattened and usually denser inside the cups.

WHERE On thin or mossy soils over limestone or lime-rich walls, outcrops and dunes.

NOTES Making observations of habitats is very important for trying to identify Pixie Cups. This species is restricted to calcareous situations. Other species will be found in different settings, so try to pay attention to clues like the presence of orange Firedots (p. 80) on nearby rocks or plants like Wild Thyme (*Thymus drucei*), which are both also found in lime-rich sites.

SIMILAR SPECIES

C. pyxidata does not form rosettes of squamules and is found in slightly more acid conditions on soils, walls and tree bases; it has smaller, rounded granules on the stalks, becoming flattened only inside the cups.

Fruticose lichens

Cladonia coccifera Lipstick Pixie Cup

HOW TO SPOT Bright red fruits along the rims of wide-flared cup-tipped stalks; the yellowish tinge of the stalks can be distinctive among more grey to brown *Cladonia* species.

DESCRIPTION Fruticose. Yellowish-grey to greenish-yellow stalks to 15mm tall, topped with wide-flared cups. Upper parts of stalks and cups with rounded, smooth granules (larger soredia with a thin cortex, so appearing compact or even smooth ), often slightly flattened within the cups, but without powdery soredia. Leafy scales (squamules) found at stalk base, rounded to slightly indented.
Reproduction: **Red apothecia** are dot-like to globe-shaped on the edges of cups.

WHERE On acid soils and peatlands, dry-stone walls and heaths.

NOTES The yellowish hue of usnic acid is important to try to distinguish as it will help you on your lichen journey, but the red apothecia with short wide cups are a giveaway here. There is some variation in the texture of the stalks and cups, apparently correlated with habitat and level of exposure, ranging from minutely squamulose at lower elevations to granular-warted at higher elevations or with greater exposure.

SIMILAR SPECIES

Look out for the distinctly blue-grey colour of **C. polydactyla** (p. 105), which also has red apothecia and can form wide cups, but these cups typically proliferate from the margins like fingers; it is often encountered on tree bases, old tree bases, deadwood and mossy soil.

Cladonia polydactyla Many-fingered Powderhorn

HOW TO SPOT One of the most abundant species on old stumps, with pale whitish-grey to bluish or greenish rough-textured stalks opening to narrow or wide cups.

DESCRIPTION Fruticose. Whitish to bluish or greenish (in shade) grey, rough-textured stalks to about 2–3cm, usually unbranched lower down, topped with irregular cups or appearing pointed when young. Cups can be narrow or expand abruptly and, when well developed, branch into finger-like extensions. Leafy scales (squamules) found at stalk bases are small and irregularly lobed, sometimes with orange pigment running down towards the base on their lower surfaces, and these sometimes have powdery soredia at their edges. ***Reproduction:*** **Red apothecia**, often rather small (), are produced at edges of cups or tips of branches. Asexual fragments ranging from tiny squamules to granules (larger soredia) with a thin cortex, which appear compact or even smooth (), to powdery soredia are variously produced, usually reducing in size farther up the stalks.

WHERE On bark and decaying conifer wood, soil and among mosses in damp and shaded woodland and heath, especially in uplands. Very common.

NOTES Its name means many-fingered, which is often an apt description. There is significant variation in the texture of the stalks and the expansion and branching of cups, but this is one of the commonest cup-forming non-yellowish *Cladonia* with red apothecia. *C. digitata* (not shown) also has red apothecia and forms narrow cups on inconspicuous stalks, but its rounded squamules are the dominant part of the lichen, up to 1cm wide, with abundant soredia on the edges of their lower surfaces.

Fruticose lichens

Cladonia floerkeana Gritty British Soldiers

HOW TO SPOT Look for grey matchsticks with red tips on rotted wood or organic soil.

DESCRIPTION Fruticose. Thallus whitish to greenish grey (in shade), stalks usually 1–2cm high and mostly unbranched, covered with smooth granules (larger soredia,), or sometimes losing these to show bare stalks, which are often darkened. ***Reproduction:*** **Red apothecia** arise at the tips; granules may also serve as asexual fragments. Granules in this species have a thin cortex, so appear smooth ( ).

WHERE On thin or acid soils, wall tops and fencepost tops and rotting wood in heaths, and on moorlands and screes.

Cladonia bellidiflora Toy Soldiers

HOW TO SPOT Tall, robust-looking, yellowish, tapered and densely squamule-covered stalks with red tips.

DESCRIPTION Fruticose. Yellowish-green stalks 3–5cm tall, covered with squamules decreasing in size from the base to the tips, making stalks appear pointed or tapered. ***Reproduction:* Red apothecia** at the tips of stalks may arise as a tiny ring of red dots ( ) or aggregate into globular irregular masses.

WHERE On thin or mossy soils on rocks in upland heaths and mountains.

NOTES As with all *Cladonia*, this species is greener in the shade. Like all yellowish-green species with usnic acid, it is more yellow in more exposed areas, as this chemical is produced as a sunscreening compound. It is rarely found without yellowish hues.

Fruticose lichens

Cladonia squamosa Dragon Horn

HOW TO SPOT Spiky *Cladonia* usually covered in squamules top to bottom.

DESCRIPTION Fruticose. Thallus whitish grey, greenish grey to brownish, stalks usually 2–5cm high and often covered with squamules appearing to peel upwards off the stalks (🍃); stalks unbranched to somewhat branched above, with branch tips pointed and sometimes opening into mostly irregular cups, which have holes down the centre, opening into the hollow stalks (🍃). New branches or apothecia may grow from cup margins.
Reproduction: **Brown apothecia**, which are small and rounded, sometimes present around cups; small squamules may also serve as dispersal units.

WHERE On nutrient-poor soil, rotting wood, old tree stumps, tree bases and mossy rocks, in heaths and in woodland. Common in the uplands and nutrient-poor parts of the lowlands.

NOTES A very common and a very variable species. Look for pointed stalks or cups with holes down the centre (see upper right image) and stalks with lots of squamules.

Cladonia coniocraea Common Powderhorn

HOW TO SPOT Probably the commonest short and spiky *Cladonia*, forming lawn-like patches on mossy bark, rotting wood or dry-stone walls.

DESCRIPTION **Fruticose.** Greenish-grey short hollow stalks to about 15mm, with pointed and usually slightly curving tips; the bottom 1–2mm of the stalks have smooth greenish cortex continuous with that on the leafy squamules from which they arise (). Leafy squamules found at stalk base are variable, incised and sometimes fringed with a soredia (powdery) covered edge, cottony white on the lower surface. *Reproduction:* **Powdery soredia**, which are so fine they are described as farinose or like flour (), are present from just above the base of the stalks to the tips.

WHERE On mossy tree bases, rotting wood or mossy dry-stone walls.

NOTES This is a relatively pollution-tolerant species and is one of the most widespread *Cladonia* across Britain and Ireland. Sometimes found with narrow cups which are not much wider than the stalks, and with a smooth layer lining the inside of the cup ().

SIMILAR SPECIES

Look out for **C. cornuta**, a slightly stouter, taller species with smooth cortex extending usually halfway up the stalk, brownish tones and pointed tips, on organic soil or rotting wood especially in Scotland.

Fruticose lichens

Cladonia furcata Many-forked Cladonia

HOW TO SPOT Loose spiky tufts of greyish green to brown on soil or mossy rocks.

DESCRIPTION Fruticose. Thallus greyish green to brown, of fine stalks usually up to 1mm wide and 2–5cm tall, forming spiky irregular to large elegant tufts; stalks often dividing evenly in 2s above, pointed and mostly smooth with cortex throughout, which is visibly broken into tiny islands (areoles,). Squamules few, at the stalk base, rounded. *Reproduction:* **Brown apothecia**, small and rounded on branch tips, can be easy to overlook and are not always present. When present, tips become stouter and less pointed.

WHERE On mossy or thin soil. A species of a wide range of habitats from heaths, dunes and coastal grasslands and even lawns to shady woodlands, mossy walls and mountains.

NOTES A very variable species, from simple spikes to branching tufts, from green in the shade to fully brown in exposed places, but usually with tips of stalks pointed and smooth texture, with very rounded squamules only at the base, and without soredia.

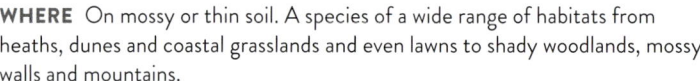

SIMILAR SPECIES

Richly branched specimens might be confused with *C. rangiformis* (p. 112), but that species has more regular and frequent branching, so branch segments appear shorter and stouter and it is found in places with slightly higher nutrients (e.g. dunes, roadsides, limestone soils).

Cladonia uncialis Heath Thorn

HOW TO SPOT Spiky yellowish tufts of broad hollow branches in heaths and on mountains.

DESCRIPTION Fruticose. Thallus yellowish (with usnic acid), up to 6cm tall, obviously hollow, branches divide by forking evenly into 2s, or sometimes into starlike tips; stalks smooth, often with marbled patterns made from flat areoles (tiny islands) of greenish to yellowish-green photosynthetic tissue contrasting with whiter structural parts of stalks (). Completely lacking leafy squamules.
Reproduction: **Probably mostly via fragmentation**, but small brown apothecia sometimes present. No soredia.

WHERE On nutrient-poor, acid soils throughout upland regions, but declining in England.

NOTES There are two forms of this species, one usually finer with long branch tips mostly splitting evenly into 2s; and one stouter, mat-forming form, with star-like branching around hollow tips. Branches can become quite wide, appearing inflated, especially on moist soils.

Cladonia rangiformis False Reindeer Moss

HOW TO SPOT Spiky dense mats often along roads and paths, often hard to spot from a distance.

DESCRIPTION Fruticose. Thallus whitish grey, greenish grey to brownish, stalks usually 2–3cm high and richly branching, by dividing evenly in 2s; stalks smooth in texture but highly patterned in colour, with round greenish patches scattered on a white background like giraffe skin (). ***Reproduction:*** **Brown apothecia** are globe-shaped, tiny and sometimes present on branch tips.

WHERE On slightly less acid soils, often along paths and roadsides, and on thin soils over limestone or in dunes.

NOTES Although this is a richly branched *Cladonia*, it does have leafy squamules at the base, giving this its common name, False Reindeer Lichen. True Reindeer Lichens (pp. 113–14) have no squamules at all.

112 The pathways

Cladonia portentosa Reindeer Lichen

HOW TO SPOT Pale bluish-green to whitish cushions of fine branches on soil in heaths, dunes, or on old stumps.

DESCRIPTION Fruticose. Thallus pale bluish green to whitish, up to 20cm wide by 12cm tall, of rather fine and richly branched mats. The largest main branches are thicker, with smaller side branches arising at all angles and further dividing until the tips, which branch mostly in 3s (🔍). The branching pattern often results in a broccoli-like shape of a mound formed of smaller mounds.
Reproduction: **Fragmentation** of branches or sometimes via tiny globe-shaped brown apothecia on branch tips.

WHERE On nutrient-poor soils or well-rotted wood in heaths and uplands.

NOTES All the richly branched *Cladonia* that arise from a crustose juvenile thallus are called Reindeer Lichens, and this is the commonest of these species in Britain and Ireland. In these, it is essential to look closely at the very ends of branches to count the tips at the very last branch points (🔍).

SIMILAR SPECIES

See p. 114 for two other Reindeer Lichens that differ by branching mostly in 4s at the tips, and which are distinctively bent at the branch ends. Branching tips consistently divided in 2s indicates *C. ciliata*.

Fruticose lichens

Cladonia arbuscula Green Reindeer Lichen

HOW TO SPOT Pale yellowish-grey cushions, with tips on each branch all curved downward and in the same direction.

DESCRIPTION Fruticose. Thallus pale yellowish grey to silvery, up to 20cm wide by 10cm tall, richly branched with clearly thicker main branches. The ends of branches bend all in the same direction and the very tips branch mostly in 3s and 4s (). Try to spot the smooth islands of cortex ( ) and the yellowish colour due to the presence of usnic acid.
Reproduction: **Fragmentation** of branches.

WHERE In open vegetation on upland and lowland heaths and dunes, sometimes on boggy soil.

NOTES Several species of Reindeer Lichens occur and are not uncommon in Scotland; getting to know them involves close inspection, especially of branch tips and texture of the cortex.

Cladonia rangiferina Grey Reindeer Lichen

HOW TO SPOT Bright whitish to creamy or brown-tipped cushions, with tips all curved downward and in the same direction per branch.

DESCRIPTION Fruticose. Thallus pale silvery to brownish or purplish due to brown branch tips, up to 20cm wide by 10cm tall, richly branched with clearly thicker main branches. On each branch, tips bend all in the same direction and branch mostly in 3s and 4s ( ). Try to spot the felty texture of the outer surfaces ( ) without islands of cortex.
Reproduction: **Fragmentation** of branches.

WHERE In montane heaths, moorland, dunes and pinewoods.

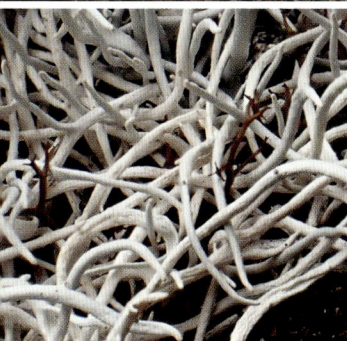

Thamnolia vermicularis Whiteworm Lichen

HOW TO SPOT White, worm-like strands, on high mountain heathlands – unmistakable!

DESCRIPTION Fruticose. Thallus chalk white, of separate cylindrical hollow strands, up to 5cm long, barely branched, pointed at the tips, sometimes forming tufts where undisturbed, but usually lying flat and separate. *Reproduction:* **Probably via fragmentation**.

WHERE On soil in high alpine heaths with low vegetation cover.

NOTES This species is a bit of an enigma, with no obvious means of sexual reproduction. However, it has a widespread distribution across Arctic and alpine regions worldwide and evidence suggests that genetic mixing (i.e. sexual reproduction) does occur – we just don't know how.

Fruticose lichens

TUFTED; BRANCHES SOLID

Stereocaulon vesuvianum Variegated Foam Lichen

HOW TO SPOT Small whitish tufts on rocks in uplands, often with 'wavy fingers' arising from tufts.

DESCRIPTION Fruticose. Thallus whitish or pale grey, tufted, with erect and sometimes wavy and tapering branches up to 3–4cm high, rarely more; branches are round and solid in cross-section, and have tiny rounded scales with darkened centres (see image below, 🔍).
Reproduction: **Globe-like clumps of soredia** (soralia) common at tips of branches, but not always present; rarely with brownish apothecia at tips.

WHERE Growing directly on silica-rich rock in the uplands; on boulders, dry-stone walls, screes, etc.

NOTES Foam Lichens produce special structures which house cyanobacteria for nitrogen fixation – cephalodia. These can be pinkish, purplish or blackish and can be smooth or irregularly spiky, and must be looked for carefully (🔍). This species rarely produces cephalodia across large parts of its distribution, as nitrogen pollution obviates the need for an internal nitrogen source. It is the only *Stereocaulon* with scales having darkened centres on its stalks.

116   The pathways

SIMILAR SPECIES

Two other common and large Foam Lichens grow directly on rock, and they differ in the shapes of the scales on the stalks: **S. dactylophyllum** (**A, B**) has tiny cylindrical scales produced all around the stalks and often has dark brown apothecia; **S. evolutum** (**C, D**) is often whiter from a distance because the scales are flatter and more densely cover the stalks; also the scales are slightly flattened and indented, like tiny paws (**D**). If you break off a single branch of *S. evolutum*, you will only be able to see the cylindrical stalks from below, as scales cover the upper sides of the stalks. Individual stalks of *S. dactylophyllum* and *S. vesuvianum* look the same when viewed from any direction, as the scales are produced on all sides.

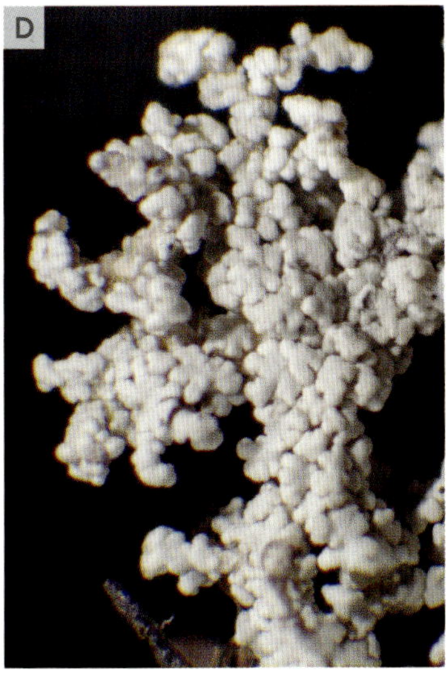

TUFTED; BRANCHES SOLID

Sphaerophorus globosus Coral Lichen

HOW TO SPOT Pale whitish-grey to orangish coral-like tufts on old oaks or birches, or on rocks.

DESCRIPTION Fruticose. Thallus pale grey, but often with brownish-orange tones especially on older branches, rarely almost completely dull orange-brown; tufted, up to 5cm tall; branches round and solid in cross-section, up to 1mm at the base with much finer cylindrical branch tips, like tiny trees with trunks and progressively smaller branches. *Reproduction:* **Apothecia** are infrequent, black and globe-shaped on the ends of branches, usually slightly drooping, with the powdery black spores loose at maturity.

WHERE On large rocks, old tree trunks and walls in uplands.

NOTES Not many lichens have loose spores at maturity. In this genus, the microscopic sacs in which the spores are produced break down early – if you touch the black spore-masses, you'll get a black smudge of spores on your finger.

SIMILAR SPECIES

S. fragilis is similar, but vegetative branches are all the same width. *Bunodophoron melanocarpum* (p. 119) is a western species, closely related and similar overall but bluer in colour and with fan-like branching; branch tips spread all in one plane at the tips, with thicker main branches slightly flattened in cross-section.

Bunodophoron melanocarpum Black-eyed Susan

HOW TO SPOT Green-grey to blue-grey tufts or extensive patches on old mossy walls or tree trunks, with longer stalks bearing drooping black masses like tiny nodding sunflowers.

DESCRIPTION Fruticose. Thallus green-grey to blue-grey, up to 4cm high, with slightly flattened branches (up to 3mm wide) laid out fan-like in layers or tiers. Shorter branchlets off main branches like little hands reaching out from the bark give rise to the nickname 'Zombie's Fingers'. Lower surface is pale to white. *Reproduction:* **Apothecia** are borne on wider and longer branches; black powdery spore-masses are downward-facing, breaking out of globe-shaped structures on branch tips.

WHERE On rocks and on bark of deciduous trees. Often found in sheltered sites among moss in woodlands. Locally common in northwest Scotland, rare elsewhere; apothecia appear to be rare outside this core distribution.

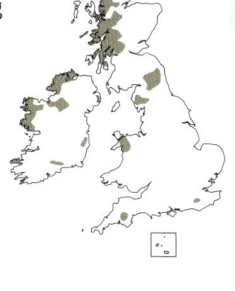

SIMILAR SPECIES

Sphaerophorus globosus (p. 118) is more widespread, and has cylindrical, not flattened, main branches, often orange-brown near their bases.

Fruticose lichens

TUFTED; BRANCHES SOLID

Rocella phycopsis Litmus

HOW TO SPOT The only pinkish-grey or purplish-grey tufted lichen in Britain and Ireland, growing on rocks and old trees in coastal areas of the southwest.

DESCRIPTION Fruticose. Thallus greyish to pinkish grey, tufted, of rounded branches arising from a holdfast, not much branched above, smooth. *Reproduction:* **Coarse soredia** produced on and near branch tips in paler, mounded clumps (soralia).

WHERE On dry sheltered spots on rocks near the coast but above the spray zone, including tombstones and church walls, very rarely on trees.

NOTES Historically, the commonly named Litmus and related Orchil lichens were collected for the dyeing industry, producing rich purple and blue colours. Today, they are too rare for collecting, so please enjoy just looking for or finding them.

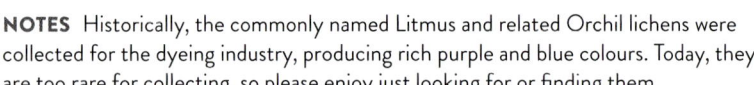

SIMILAR SPECIES

R. fuciformis (Orchil) is similar, but with flatter branches and even rarer than *R. phycopsis*. The coastal species of Sea Ivory *Ramalina siliquosa* and *R. cuspidata* (opposite) are similar, but have no soredia, and are usually greenish to whitish.

TUFTED; BRANCHES FLATTENED

BD 1

Ramalina siliquosa Sea Ivory

HOW TO SPOT Individual tufts or whole lawns of pale greenish flattened branches on rocks, cliffs and dry-stone walls near the sea.

DESCRIPTION **Fruticose.** Thallus pale greenish, mostly up to about 6cm long, forming individual tufts each attached by a central holdfast; multiple mostly unbranched strap-like lobes, 2–5mm wide, but sometimes much longer and up to almost 1cm wide, especially on vertical rock faces. Tips of branches with paler bumpy areas housing asexual spore-producing structures (pycnidia,), these with white pores. *Reproduction:* **Apothecia** the same colour as the branches with flat to convex discs; sometimes inconspicuous, they are formed on tips of branches, up to about 6mm wide.

WHERE On seaside rocks, cliffs and walls, and sometimes inland on gravestones, tall stone monuments and boulders. Sometimes forming huge colonies covering entire cliffs.

SIMILAR SPECIES

R. cuspidata has narrower and less flattened branches which are blackish at the base, and they have black bumps (pycnidia) on branches instead of whitish ones. *R. cuspidata* is restricted to coastal locations.

Fruticose lichens   121

TUFTED; BRANCHES FLATTENED

BD 1

Ramalina farinacea Shaggy Strap Lichen

HOW TO SPOT Pale green short tufts of strap-shaped branches on bark, usually hanging down.

DESCRIPTION Fruticose. Thallus pale green on all surfaces; flat, strap-shaped branches up to 6cm long and 3mm wide grow from a single holdfast. *Reproduction:* **Fine powdery soredia** formed in pale oval patches (soralia) on the edges of branches ( ); individual soredia are very small in size, like fine flour, and difficult to distinguish even with a hand lens.

WHERE On bark of deciduous shrubs and trees and common throughout, but not on acidic bark. Rarely on rock.

NOTES Like all *Ramalina* species, the texture is tough like cartilage, and thalli are the same colour throughout with strap-shaped branches. This is an indicator species of medium nutrient levels and medium-aged habitats and is often found with Whitewash Lichen *Phlyctis argena*.

Phlyctis argena

122   The pathways

SIMILAR SPECIES

Evernia prunastri (p. 130) may appear similar, but its branches have irregular ridges and irregular patches of soredia rather than the neat round oval soralia of *R. farinacea*, and its young branches are white on the lower surface. *R. fastigiata* has short-tufted upright thalli (**A**) with many branches tipped with apothecia (**B**), leading to its occasional nickname 'Shrek's Ears'. *R. fraxinea* (**C**) has fewer irregularly ridged, wide, flat branches up to 1–3cm wide with apothecia on and near the edges of branches. *R. calicaris* is a western species with channelled branches (slightly U-shaped in cross-section, **D**) and apothecia on the lobe edges just before the tips, often drooping from tree branches (**E**).

TUFTED; BRANCHES FLATTENED

Flavocetraria nivalis Crinkled Snow Lichen

HOW TO SPOT Most pale yellowish lichens at high altitude in heathlands are Reindeer Lichens in the genus *Cladonia* (pp. 113–14), with narrow cylindrical branches – so keep an eye out for these denser patches of pale yellow, made of erect, wide and flattened, crinkled lobes in acid heathlands at high altitudes.

DESCRIPTION Fruticose. Pale yellowish, erect tufts up to 4–5cm tall with wide strap-shaped branches, slightly channelled and irregularly wrinkled and furrowed, with indented tips.
Reproduction: Probably mainly by vegetative fragmentation in Britain and Ireland; apothecia very rare on tips of branches.

WHERE On soil above 900m in nutrient-poor alpine heath and mossy vegetation.

NOTES A very rare species in Britain and Ireland, found only on the high heaths of the Cairngorms, but much more common in other Arctic or alpine habitats in Europe.

124   The pathways

FOLIOSE LICHENS

We help you find the best match for your foliose lichen by a process of elimination. Separating foliose lichens by a combination of colours, lobe shapes and lower surfaces shown in the Quick Guide of thumbnail images on the next page helps narrow your options. When you find a best match using the thumbnails, go to the page number for the Species Description.

1. **Bold colours**
   ▪▪ Orange, yellow and bright green when wet are very distinctive. They are placed in two groups with corresponding coloured tabs.

2. **Muted colours and brown/black**
   ▪ For most foliose lichens you encounter, it is important to look at individual lobes and pay attention to habitats. For most of the pale greyish, greenish to brown lichens there is a wide variation in colour, and lobe shapes and lower surfaces are important. Try to recognise these groups:

   - **Strap-shaped** Two foliose lichens that may appear fruticose are distinctive: the strap-shaped *Evernia prunastri* and *Pseudevernia furfuracea*. In these species, each branch or lobe is different on the top and bottom, making them true foliose lichens.
   - **Jelly Lichens** These dark-brown to blackish lichens look like dark patches in mossy places or on calcareous rocks (Baker's Dozen, p. 78).
   - **Lower surface felty or with pores or veins** Mostly large (lobes usually over 1cm). All of these species live in moist habitats or microhabitats like temperate rainforests or mossy tree bases, have cyanobacterial photobionts, and turn darker when wet. Think Big Browns and Pelts (Baker's Dozen, pp. 74–77).
   - **Attached centrally, only on rock** A few lichens are attached only centrally, by something like an umbilical cord (thus the name *Umbilicaria*). This is easy to tell when the lichens are made of a single, more-or-less circular lobe. Take the time to look around, and see if you can spot any young thalli, where the attachment is clearer.
   - **Medium lobes, up to 1cm wide** A large number of foliose lichens have slightly shiny upper surfaces and lobes usually 2–5mm (up to 1cm) wide. Many of them are related to the genus *Parmelia* (Baker's Dozen, p. 70). See the next page for thumbnails of some of these Parmelia-group lichens.
   - **Narrow lobes up to 2mm wide** Lobes look narrow, because they appear longer than wide, or are actually really narrow and less than about 2mm; most of these have matt surfaces. These are mostly small lichens with thalli less than about 5cm across. Most belong in the Rosettes, Frosts and Fringes group (Baker's Dozen, p. 72).

TOP TIP Pale greyish lichens especially can be hard to separate out when they're growing close together. Take time to look at the individual lobes carefully.

QUICK GUIDE

FOLIOSE LICHENS

BRIGHT GREEN (WET)

Melanelixia (p. 133)

Phaeophyscia and *Physconia* (pp. 132, 177)

Lobaria (p. 136)

Ricasolia (p. 137)

Peltigera (p. 138)

Anaptychia (p. 176)

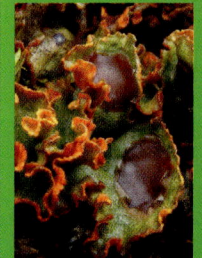
Solorina (p. 134)

Lasallia (p. 135)

MUTED COLOURS AND BROWN/BLACK

LOBES MEDIUM, UP TO 1cm WIDE

Parmelia group (p. 128)   *Nephroma* (p. 158)

LOBES NARROW, UP TO 2mm WIDE

Anaptychia (p. 176)   *Phaeophyscia* (p. 132)

ATTACHED CENTRALLY, ONLY ON ROCK

Umbilicaria (p. 155)   *Lasallia* (p. 135)

LOWER SURFACE FELTY OR WITH PORES OR VEINS, MOSTLY LARGE

Sticta (pp. 149–50)   *Pseudocyphellaria* (p. 151)

Peltigera (p. 145)   *Ricasolia* (p. 137)

126 The pathways

LOBES MEDIUM, UP TO 1cm WIDE

Platismatia (p. 172)

Parmelia group (p. 128)

Hypogymnia (p. 173)

LOBES NARROW, UP TO 2mm WIDE

Arctoparmelia (p. 175)

Physcia & *Physconia* (pp. 177–80)

BRANCHES STRAP-SHAPED

Evernia (p. 130)

ATTACHED CENTRALLY, ONLY ON ROCK

Dermatocarpon (p. 157)

Umbilicaria (p. 156)

Pseudevernia (p. 131)

LOWER SURFACE FELTY OR WITH PORES OR VEINS, MOSTLY LARGE

Lobarina (p. 148)

Peltigera (pp. 145–46)

Pectenia and *Pannaria* (pp. 153–54, 181)

Jelly-like texture when wet, (pp. 142–44)

YELLOW TO ORANGE

Xanthoria & *Polycauliona* (p. 139)

Candelaria & *Polycauliona* (p. 141)

QUICK GUIDE
PARMELIA GROUP

YELLOWISH

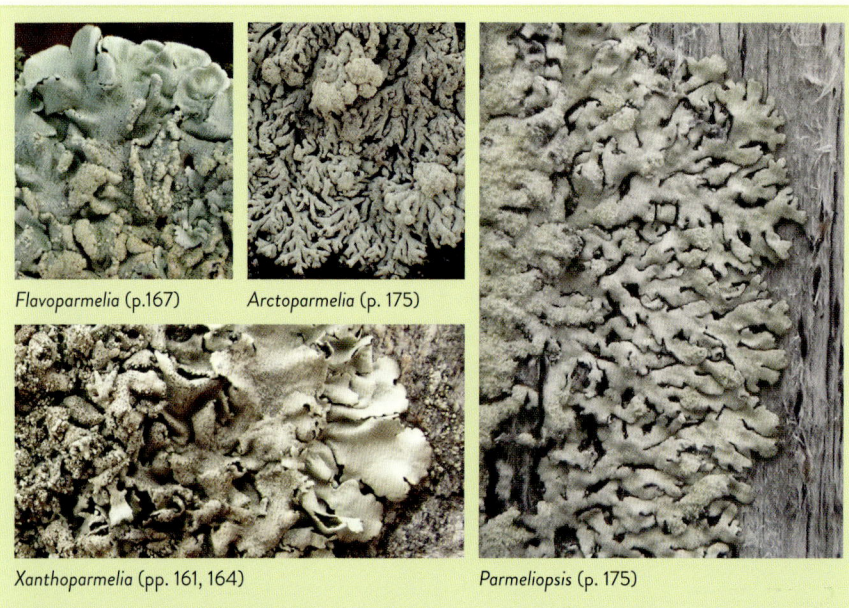

Flavoparmelia (p.167)   Arctoparmelia (p. 175)

Xanthoparmelia (pp. 161, 164)   Parmeliopsis (p. 175)

KHAKI, GREENISH, BROWN TO BLACK

Parmelia (p. 162)   Melanelixia (pp. 133, 159)

Melanelixia (p. 160)   Xanthoparmelia (p. 161)

PALE WHITISH OR GREY TO GREENISH

LOBES INFLATED, HOLLOW

Hypogymnia (p. 173)

Menegazzia (p. 174)

LOBES *WITH* WHITE DOTS, LINES OR CRACKS ON UPPER SIDE (🔍)

Parmelia (pp. 162, 163, 165)

Punctelia (p. 166)

Platismatia (p. 172)

LOBES *WITHOUT* WHITE DOTS, LINES OR CRACKS (PSEUDOCYPHELLAE) ON UPPER SIDE (🔍)

Parmotrema (p. 171)

Flavoparmelia (p. 167)

Hypotrachyna (pp. 169–70)

Xanthoparmelia (p. 164)

Parmelina (p. 168)

Evernia prunastri Oak Moss

HOW TO SPOT Look for pale green bushy to drooping tufts with wide strap-shaped lobes on small branches and twigs, and on tree trunks in less polluted areas.

DESCRIPTION Foliose. Thallus pale green with lobes to 2–3mm wide, 3–5cm long, and branching by dividing evenly in 2s; lobes have an irregular network of ridges and hollows. Check young lobes which are flat, neatly branched and clearly differentiated with the upper surface green, contrasting with the chalk-white lower surfaces. Older lobes may curl and twist, with greenish patches on upper and lower surfaces. *Reproduction:* **Powdery soredia** are produced in inconspicuous irregular clusters (soralia) and found on the ridges and lobe edges ( ); it is unusual to find its bowl-like, chestnut-brown apothecia.

WHERE On bark, especially common on deciduous trees and shrubs, but tolerant of a wide range of microsites, except in areas with very high air pollution.

NOTES This is the only species of *Evernia* in Britain and Ireland, and is very common throughout. It looks fruticose because it's bushy, but is actually foliose because its surfaces are different top and bottom. Widespread in Europe and North America and was once harvested commercially for the perfume industry, for its slightly sweet, woody notes and fixative properties.

SIMILAR SPECIES

Ramalina farinacea (p. 122) is a similar colour and occurs in similar habitats, but its branches are the same pale green all the way around (it's a fruticose lichen), without an upper or lower surface. *Pseudevernia furfuracea* (opposite) has similar branching in 2s, but is pale whitish to dark lead grey on the upper surface, black on its lower surface, and has abundant, long isidia. It prefers more acid conditions, especially conifer twigs and bark.

Pseudevernia furfuracea Antler Lichen

HOW TO SPOT Look for grey and black tufts of forking branches on conifer twigs and fenceposts.

DESCRIPTION Foliose. Thallus pale to medium grey above with a black lower surface, tufted, to 10cm, of strap-like and forking branches up to 5mm wide, channelled (slightly U-shaped in cross-section). *Reproduction:* **Tiny finger-like isidia** abundant, long and cylindrical on surfaces and edges of lobes ( ), sometimes expanding into flattened lobes.

WHERE On twigs and branches of acid-barked trees, especially conifers, but also birches, and on wooden fences and sometimes silica-rich rocks.

NOTES This lichen is rather three-dimensional, but is still considered foliose, as the algal layer is only on the upper side. Many identification keys will treat it in both the fruticose and foliose sections. You'll often find this growing with *Hypogymnia*, *Usnea* and *Platismatia*.

SIMILAR SPECIES

Evernia prunastri (opposite) has a similar growth form, but is green and white rather than grey and black. The rare and declining *Anaptychia ciliaris* (p. 176) occurs on deciduous trees, especially in the south, and has long and obvious cilia on branches rather than isidia.

Phaeophyscia orbicularis Mealy Shadow Lichen

HOW TO SPOT Gregarious colonies of flat, closely pressed tiny thalli, often inconspicuous unless wet or on a contrasting background. Usually spotted with a lens when looking for something else!

DESCRIPTION Foliose. Thallus green when wet, otherwise grey to dark grey to brownish grey, lobes less than 1mm, thallus usually less than 2cm, forming neat rosettes. Dense black and white rhizines often project from underneath lobe tips ( ). ***Reproduction:* Powdery soredia** produced in small round or oval patches (soralia) on the surface of lobes ( ).

WHERE On nutrient-enriched bark, also concrete, signs and other human-made materials.

NOTES Often found with *Physconia* (also green when wet, see p. 177), *Physcia* species and *Xanthoria parietina* (p. 139).

SIMILAR SPECIES Look out for the tiny greenish thalli up to 1.5cm of ***Hyperphyscia adglutinata***, which does not have projecting rhizines and is pressed incredibly tightly to its substrate. It is found particularly in the south, and its thalli sometimes coalesce to form large patches.

Melanelixia glabratula Polished Camouflage Lichen

HOW TO SPOT Usually not easy to spot unless wet. Greenish-brown to brown rosettes on trunks of hardwood trees, often well camouflaged against the bark or rarely standing out against grey leafy lichens or pale bark.

DESCRIPTION Foliose. Thallus olive green, or brownish, greener when wet, 3–8cm, forming rounded rosettes, quite flat against bark, with slightly incised and irregular tips. The surfaces of lobe tips are shiny ( ). ***Reproduction:*** **Fine, cylindrical isidia** form individually (check young lobes where they are sparser) and can become dense and branched in older parts of the thallus ( ).

WHERE On trunks of broadleaved trees, especially on vertical bark.

NOTES A number of brownish to olive-green '*Parmelia*' group lichens are called Camouflage Lichens because they are often tricky to spot. They can be learned individually by their reproductive structures. When you see one, ask yourself how it reproduces, and you will start to see the differences.

SIMILAR SPECIES

Get in the habit of using your lens to tell Camouflage Lichens apart. *M. subaurifera* (p. 159) is often browner, with soredia arising in clumps and more often grows on twigs and branches.

Solorina crocea Chocolate Chip Lichen

HOW TO SPOT Unmistakable when in high-elevation mountain heaths, where the bright orange colour on soil stands out.

DESCRIPTION Foliose. Thallus greenish (wet), brownish (dry), up to 12cm, rosette-forming, lobes up to 1cm wide, with slightly ruffled edges, where the bright orange cottony lower surfaces are visible. *Reproduction:* **Apothecia** large, flat and chestnut brown, like chocolate chips in a cookie, arranged singly in the centres of the upper surfaces of lobes.

WHERE On soil in moist and open mountain heaths above 900m, often with slightly richer soils.

NOTES Look out for *Thamnolia vermicularis* (p. 115) and *Stereocaulon* species (p. 116) as these are found in similar habitats.

SIMILAR SPECIES

S. saccata has rounded, individual lobes with dark apothecia in their centres and is often found as small groups of lobes. It lacks orange lower surfaces and is usually on bare lime-rich soils in grassland or near limestone outcrops.

Lasallia pustulata Poppadom Lichen

HOW TO SPOT Messy groups of loose leaves on large boulders in the west, with unmistakable bumpy texture up close – greenish when wet, grey when dry.

DESCRIPTION Foliose. Thallus greenish (wet), brownish grey to pale grey (dry), up to 15cm but often smaller; plate-like and attached only centrally, so appearing loose, with ragged edges; upper surface finely roughened with crystalline coating (), and with abundant large, rounded swollen areas like the bubbly surface of a poppadom. The lower surface has hollows corresponding to the swollen areas above. ***Reproduction:*** **Isidia** densely branching and formed in dense dark patches nearer the edges of the thallus (). Apothecia rare, blackish and smooth, round.

WHERE On the tops and upper sides of large boulders in the west, especially in nutrient-enriched sites.

NOTES *Lasallia* and *Umbilicaria* (pp. 155–56) are called Rock Tripe Lichens and have been used for survival food, but they require extensive washing and cooking. Unlike most Rock Tripes in Britain and Ireland, this one is often found without other related species nearby, as most of the other species are sensitive to nutrient enrichment.

Foliose lichens   135

dry   isidia

Lobaria pulmonaria Lungwort

HOW TO SPOT Lettuce-sized lobes lifting from tree branches and trunks, and ornamented with a network of conspicuous raised ridges – unmistakable.

DESCRIPTION Foliose. Thallus bright green (wet) to pale brownish (dry) above, up to 30cm across, with individual lobes growing away from the bark, each up to 20cm long and up to 8cm wide, with a conspicuous and unmistakable pattern of ridges and rounded to angular hollows. The lower surface is pale cream and covered with fine felty hair (tomentum) within a network of depressions, the inverse of the ridges on the upper side (p. 77, **H**); the tomentum becomes darker towards inner older parts. *Reproduction:* **Apothecia** reddish brown with a paler margin, up to 4mm wide. Mature lobes also produce coarse soredia and isidia on ridges.

WHERE On deciduous trees and shrubs in ancient damp woodlands in the east and more widespread in the west, found on old walls and even garden shrubs.

NOTES This lichen is an indicator of ancient woodland in many parts of its extensive range and is endangered in parts of continental Europe. It gives its name to a whole community of lichens that occur in mature woodland in oceanic climates, the Lobarion. Check the section on 'Big Browns – Species of Wet Woodlands' (p. 76) for species you are likely to see nearby.

Ricasolia virens (Lobaria virens) Green Satin Lichen

HOW TO SPOT Especially conspicuous when wet, as bright green extensive patches the size of dinner plates, dotted with brown discs.

DESCRIPTION Foliose. Thallus bright green (wet), duller and paler greenish to brown (dry), up to 20cm; lobes 5–10mm wide, attached along their length to bark, radiating and branching irregularly to create characteristically rounded rosettes or extensive patches, with rounded tips and wrinkled undulations centrally ( ).
Reproduction: **Apothecia** abundant centrally, prominent and donut-shaped when immature, becoming even more conspicuous, and orange-brown, with paler margins, up to 3mm.

WHERE In old woods, particularly on trees with mildly acidic to neutral pH bark (elm, ash, hazel) in addition to beech and oak, but also on rocks in sheltered western woods or sea-cliffs. Rare and declining in England but locally common in western Scotland and far western Ireland.

NOTES Like many Lobarion lichens which may be locally common in western Scotland, this species is restricted to ancient woodlands in drier settings.

SIMILAR SPECIES

The similarly large, rosette-forming *Ricasolia amplissima* is pale whitish grey when dry, and with notched lobes, and without apothecia, but instead with dark, tufted cephalodia (cyanobacteria-containing structures).

Foliose lichens

Peltigera leucophlebia Ruffled Freckle Pelt

HOW TO SPOT Bright green, ruffled and dark-freckled, large-lobed lichen near damp mossy rocks.

DESCRIPTION Foliose. Thallus bright green (wet), greyish green to brownish (dry), pale creamy on the lower surface, with dark brown veins, up to 20cm or more. Lobes up to 4cm wide, rounded at the tips, with curled and ruffled edges often showing their creamy lower surfaces. Upper surface speckled with small (just over 1mm), irregular greyish-purple cyanobacteria-containing structures (cephalodia), which cannot be easily picked off. *Reproduction:* Apothecia infrequent, on tips of erect and curved lobes, reddish brown.

WHERE On soil among mosses and rocks in damp lime-rich habitats.

SIMILAR SPECIES

P. britannica here shown dry is similar but has wider, simpler and unruffled lobe edges, with cephalodia that are easily detached. It prefers more acid mossy rocks, walls or tree bases.

Xanthoria parietina Common Sunburst Lichen

HOW TO SPOT Bright yellowish or orange rosettes on all sorts of surfaces, bark or rock, across most of the lowlands.

DESCRIPTION Foliose. Thallus usually bright yellow to orange but greenish grey in shade, up to 15cm, but often smaller, often with inner older parts falling away. Lobe tips mostly pressed closely to the surface they are growing on, individually wide-spreading, papery-textured and overlapping, up to 5–7mm wide at tips. *Reproduction:* **Apothecia** are abundant centrally, orange and cup-shaped, slightly raised from the thallus.

WHERE On nutrient-enriched trees, rocks, roofs, walls, paintwork and even long-parked cars in urban areas and parks, farmland, along road networks, and abundant on rocky shores throughout. Widespread and increasing in nutrient-enriched places.

NOTES The orange pigment is produced in response to light, so shaded thalli are greenish grey. Often found with lichens tolerant of nutrient inputs, including *Physcia* and Firedots.

SIMILAR SPECIES

X. calcicola (**A**) is similar but has thicker lobes with knobbly isidia covering mature inner parts and few to no apothecia; it is a more southern species of lime-rich or nutrient-enriched microhabitats, like concrete, mortar or bird-perch rocks, and coastal rocks in the north.

X. aureola (**B**) is coastal with more elongate lobes, often separated and more squared at the tips, giving it the nickname Cornflakes. All three of these species can be found together on coastal rocks. See also *Polycauliona candelaria* (p. 141).

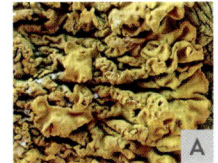

Polycauliona polycarpa (Xanthoria polycarpa)
Pincushion Sunburst Lichen

HOW TO SPOT Tiny but bright yellowish or orange cushions of apothecia on top of inconspicuous lobes – usually found when searching for something else.

DESCRIPTION Foliose. Thallus usually bright yellow to orange or greyish in shade, up to 2cm. Lobes convex and finger-like, pressed closely to the surface they are attached to and narrow, only about 1mm wide and often barely extending beyond the central collection of apothecia. *Reproduction:* **Apothecia** are abundant centrally, orange and cup-shaped, slightly raised from the thallus and usually more conspicuous than the thallus itself.

WHERE On rocks and on twigs especially where branches diverge. Widespread but decreasing.

NOTES The orange pigment is produced in response to light, so shaded thalli are greenish grey. Often found with lichens tolerant of nutrient inputs, including *Xanthoria parietina*, *Physcia* and *Physconia*, along with various Firedots on rock.

SIMILAR SPECIES

Rusavskia elegans (formerly *Xanthoria*) also has narrow convex lobes but is a deep reddish-orange species up to about 5cm diameter, with separate lobe tips that can be peeled off the surface it is growing on.

Polycauliona candelaria (*Xanthoria candelaria*)
Shrubby Sunburst Lichen

HOW TO SPOT Irregular patches of orange or yellow on bark or rocks with tiny but upright lobes.

DESCRIPTION Foliose or appearing minutely fruticose (). Thallus lemon-yellow to yellow-orange to dark orange on both surfaces, forming loose cushions up to about 2cm or spreading more irregularly and widely. Tiny upright lobes to about 1mm wide, appearing delicately incised at their tips (). ***Reproduction:*** **Tiny portions bud off edges of lobes** () and apothecia are sometimes formed.

WHERE On nutrient-enriched rocks, tree bark, tops of fenceposts and gravestones.

NOTES A few closely related species are included under this name that vary in shape of lobes, colours and degree of differentiation between upper and lower surfaces.

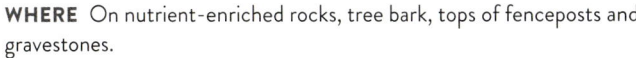

SIMILAR SPECIES

Candelaria concolor is a tiny, yellow foliose lichen on bark, fences, rocks and walls, also forming cushions of ascending lobes, but these are very narrow (0.5mm or less) and lobes lie loosely, but mostly flat, against the surfaces they grow on; they are white on the lower surfaces ().

Scytinium gelatinosum (*Leptogium gelatinosum*)
Rose-petalled Jellyskin

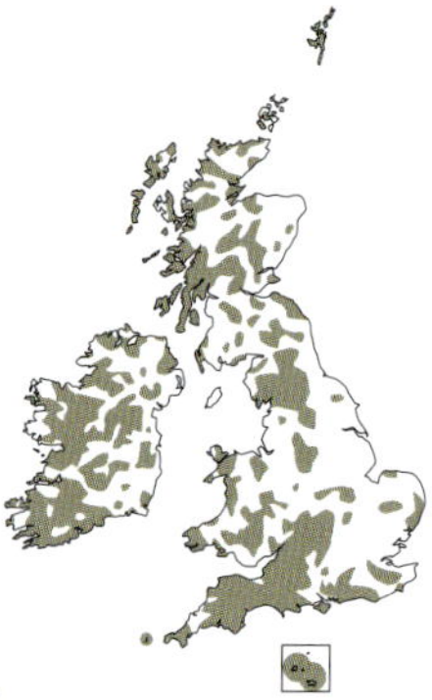

HOW TO SPOT Look for a dark patch among mosses on old walls or trees, then check with your lens – it's probably a Jelly Lichen.

DESCRIPTION Foliose. Thallus dark reddish brown to brown-black up to 8cm, of very thin lobes, rounded or divided at tips, up to 3–5mm wide, usually overlapping, distinctly wrinkled above when dry (), but not much swollen when wet. *Reproduction:* Apothecia to 2mm, with a reddish-brown disc and slightly raised margin ().

WHERE On walls, limestone soils, dunes, among mosses in lime-rich situations, only rarely at the bases of trees.

NOTES Jelly Lichens (p. 78) are dark in colour because they contain cyanobacteria rather than green algal photobionts. This means they are able to fix nitrogen as well as carbon, and they grow in places with plenty of liquid water. *Scytinium* Jellyskin Lichens are thin, without much swelling when wet and slightly shiny when dry ().

SIMILAR SPECIES

S. lichenoides is similar, but has fine, isidia-like outgrowths () on lobe tips; it is common on mossy bases of trees and walls.

Lathagrium auriforme (Collema auriforme) Eared Jelly

HOW TO SPOT Distinctive blackish green and jelly-like look when wet, thinner when dry – search for black rosettes on lime-rich rocks, often among mosses in damp shaded sites and wet shady crevices.

DESCRIPTION Foliose. Thallus dark green-brown to blackish, forming cushions or radiating in rosettes, lobes up to 1cm wide, ear-like (auriform) in shape, very swollen when wet and matt when dry. *Reproduction:* **Isidia** are globose or bead-like (), occurring in patches sometimes covering the centre of the thallus.

WHERE On hard lime-rich rocks; common on marble tombstones. Sometimes found on ground in chalk or limestone grasslands.

NOTES Jelly Lichens (p. 78) are dark colours because they have cyanobacterial photobionts, and it is important to try to identify them when dry, when you can see their textures clearly. *L. auriforme* has tiny wrinkles when dry (). Watch out for dark, leafy lobes of free-living colonies of *Nostoc*, a cyanobacteria living independently, often on old pavements, tracks or lime-rich soils. It does not have any reproductive structures of lichens (), but does have large, contorted, leafy lobes and is gelatinous when wet.

Foliose lichens 143

wet | dry

Enchylium tenax (Collema tenax) Soil Jelly Lichen

HOW TO SPOT Look for little black patches in human-made environments, often underfoot.

DESCRIPTION Foliose. Thallus blackish (wet) to dark or reddish brown (dry) to 4cm, of small erect lobes centrally and with smaller radiating lobes on the edges. Lobes clearly crumpled when dry (🔍) and very swollen when wet. *Reproduction:* **Apothecia** with thick margins sometimes on the tips of lobes.

WHERE On soil in nutrient-rich, especially human-made environments such as between bricks, paving stones or on crumbling mortar.

NOTES Like all Jelly Lichens (p. 78), this species is darker when wet because of its cyanobacterial photobiont. When wet, it appears as densely packed, black shiny fingers all pointing up, with radiating stubby lobes on the edges (🔍).

SIMILAR SPECIES

Blennothallia crispa is also common on mortar and paths but is a thinner-textured Jelly Lichen with tiny ear-like scales (🔍) covering the inner parts; these become brittle and fragment for reproduction and dispersal.

Peltigera hymenina Smooth Pelt Lichen

HOW TO SPOT Ruffled rosettes with upturned edges in damp mossy places.

DESCRIPTION Foliose. Thallus grey to brown (dry) to blackish grey or greenish black (wet), up to 15cm, with smooth upturned and curled lobes up to 2cm wide. Lower surface cottony white, with a network of slightly raised but flattened pale brown veins and simple rhizines with tufted tips. *Reproduction:* **Apothecia** are frequent, and longer than wide, produced on upturned lobes whose margins roll together to expose the reddish-brown spore-bearing surface.

WHERE On mossy soil, rocks and tree bases in nutrient-poor situations (see habitat photo p. 28).

SIMILAR SPECIES

Peltigera rufescens also has ruffled and upturned thallus edges, but is usually more pressed to the soil or moss and is usually browner to reddish brown when wet, with a pale dusty white bloom sometimes visible on dry lobes. It has dark, tufted rhizines and raised veins which are darker in the centre of the thallus.

lower surface

Foliose lichens

Peltigera membranacea Membranous Pelt Lichen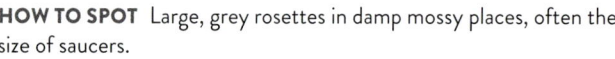

HOW TO SPOT Large, grey rosettes in damp mossy places, often the size of saucers.

DESCRIPTION Foliose. Thallus pale grey (dry) to dark grey or greenish brown (wet), up to 30cm, with neatly radiating, rounded lobes up to 3cm wide. Upper surfaces of lobes have convex, raised areas and fine felty hairs at the tips (; see Notes below). Lower surface cottony white, with a network of prominent white veins and long, dagger-like, hairy, but unbranched and separate rhizines.
Reproduction: **Apothecia** produced on upturned lobes whose margins roll together to expose the reddish-brown spore-bearing surface.

WHERE On mossy soil, rocks and tree bases in nutrient-poor situations.

NOTES The long, white rhizines look like canine teeth of dogs; for this reason, it was once used to treat dog bites, but we don't recommend that! To see the fine felty hairs on the upper surface, squeeze a lobe between your fingers to press out moisture on a wet sample – the hairs will then lift out of the surface film of water.

For the Pelt Lichens, it is important to look at the lower surfaces. In this species, the raised pale veins and dagger-like rhizines are characteristic.

rhizines and veins

SIMILAR SPECIES

P. praetextata (**A**) is a similar western and northern species but has diagnostic tiny lobes along edges and often on damage or cracks in the upper surface (**B**) along with long simple rhizines on the lower surface. *P. horizontalis* (**C**) is a western, old woodland indicator species, blackish brown, with smooth and neatly rounded wide lobes up to 3cm, and oval apothecia held horizontally (**D**).

Foliose lichens

lower surface

Lobarina scrobiculata Textured Lungwort

HOW TO SPOT Wide, grey round-ended lobes very loosely attached on old broadleaved trees, forming patches often over 15cm.

DESCRIPTION Foliose. Thallus greyish blue (wet), yellowish grey (dry), up to 15cm (or more) in size, loosely attached and growing away from substrate; rounded, little-branched lobes are 1–3cm wide and pitted with large, rounded and irregular depressions. Lower surface creamy to pale brown with a covering of even short hairs ( ) alternating with white hairless patches that correspond to the depressions on the upper surface. *Reproduction:* **Powdery soredia** are coarse in grey, mounded clusters (soralia), spreading along ridges on the lobes and lobe edges ( ). Apothecia are reddish brown with pale margins, up to about 2mm, infrequent.

WHERE On old broadleaved trees, including wayside trees; sometimes in coastal areas, on mossy rocks.

NOTES This species has declined markedly in many parts of its range, particularly in England, but also more widely in continental Europe. It is affectionately known as Lob Scrob.

SIMILAR SPECIES

Ensure you have a good look at the lower surface to exclude Pelt Lichens (*Peltigera*), which have a network of raised veins and prominent, usually separate, rhizines.

lower surface

Sticta limbata Powdered Moon Lichen

HOW TO SPOT Few-lobed, greyish-brown lichen protruding well off the trunks and branches of trees in western woods, reminiscent of patches of suede.

DESCRIPTION Foliose. Thallus greyish brown (dry) to rich chocolate brown (wet), up to 10cm across, with wide lobes to 3cm, often upturned; lower surface creamy to pale brown and evenly short-hairy with neatly rounded paler pores, each with a tiny neat rim of tissue outlining it (cyphellae,). ***Reproduction:* Powdery soredia** are formed in paler bluish-grey patches (soralia), especially on lobe edges and spreading inward.

WHERE On bark and sometimes mossy rocks in moist western woods, including both trunks and branches. It is tolerant of open conditions and canopy gaps.

NOTES This is the only Moon Lichen with soredia in Britain and Ireland, but like all *Sticta* species it has a distinctive fishy smell when damp – you may need to rub it a little to detect this: stinky *Sticta*!

SIMILAR SPECIES

Several other brownish to greyish lichens with soredia are found in western woods – always check their lower surface first (see Big Browns, p. 76). ***Peltigera collina*** has soredia on its upturned and ruffled lobe margins (**A**); it tends to be rosette-forming, with lobes up to about 1cm wide. Its lower surface has flattened pale to dark brown veins and short-tufted rhizines (**B**).

lower surface | isidia

Sticta sylvatica Bay Moon Lichen

HOW TO SPOT Large, dark blackish-brown scalloped lobes extending outwards from mossy trees and rocks in the west.

DESCRIPTION Foliose. Thallus blackish brown and somewhat glossy when dry, large, to 15cm across with lobes up to 3cm, with wide-rounded and upturned lobes. Lobes are branched at the tips into broad scallops undulating in uneven ridges. Lower surface pale to dark brown, darker towards the centre and evenly short-hairy, with pale, neat, sunken pores edged with a thin membrane (cyphellae, 🔍). *Reproduction:* **Isidia** long-cylindrical, much-branched (🔍), and grouped in clusters, arising on ridges.

WHERE On trunks and mossy rocks in moist western woodlands.

NOTES Like all Moon Lichens, this has a distinctive fishy smell when damp, thus nickname Stinky *Sticta*. You may need to rub the thallus a little to detect this. 'Bay' in the common name refers to the blackish brown colour.

SIMILAR SPECIES

As with all brown lichens in western woods, check the lower surface. In *Sticta*, look for neat pores with a delicate membrane edge. The Common Moon Lichen **S. fuliginosa** has unbranched, rounded lobes and a downturned edge.

lower surface

Pseudocyphellaria citrina Golden Specklebelly

HOW TO SPOT Large brown loosely attached lichen with wrinkled lobes and golden yellow spots and patches.

DESCRIPTION Foliose. Thallus milk-chocolate brown (dry) or dark brown to slate grey (wet), up to 15cm or more, irregularly spreading; lobes up to about 15mm wide, with scalloped edges and irregular wrinkles. *Reproduction:* **Powdery soredia** are bright golden yellow (dry) in rounded clusters (soralia), fringing the edges and dotted across the surface of lobes, especially following irregular ridges and wrinkles. When wet, check for the yellow colour with your hand lens.

WHERE On trees with mildly acidic to neutral bark (ash, hazel) and mossy rocks; widespread, but locally distributed – in the rainforest zone of Scotland and Ireland and one site in Cornwall.

NOTES Completely unmistakable when dry. Look out for it when you see other members of the Big Browns (p. 76). The lower surface is covered in fine, felty, brown hairs and has dot-like yellow patches ().

Foliose lichens 151

Sticta canariensis Canary Moon Lichen

HOW TO SPOT This is one to go on safari for – it's quite rare, but worth that special trip! If you're lucky you'll spot a brownish-black (wet) leafy lichen with delicately incised lobe tips and brilliant green lobes growing directly from it. When dry, it can be very tricky to find.

DESCRIPTION Foliose. Three forms: 1. Thallus dull brown to grey, delicately incised at the margins of the rounded lobes, in irregular patches up to 12cm or more; bluish-brown lobes are up to 2cm wide and have a fine paler net-like pattern on the upper surface (⊙). 2. The brown-grey form may give rise to slightly square-tipped green lobes, with forking branches. 3. Thallus of green forking lobes to 1.5cm wide, with subtle ridges on the upper surface. All morphs have fine creamy to dark brown short hairs and scattered neat pores (⊙, cyphellae) on the lower surface. *Reproduction:* **Isidia** are produced by the brown morph; these are fine and richly branching from the centres and edges of lobes. The green morph occasionally produces reddish apothecia.

WHERE In very moist and usually shaded mossy places on trees and rocks in oceanic settings – in the far west or near the coast. The green form is much rarer.

NOTES The different forms are caused by a single fungus – *Sticta canariensis* – having either a cyanobacterial or a green algal partner, and sometimes both at the same time. Brown or grey lobes indicate a cyanobacterial photobiont, and green ones a green alga. It is not known why the shapes of the lobes are so different with the different photobionts, but it's fascinating that this single fungus can form associations across these vastly different domains of life. Several relatives of *Sticta* have the ability to produce these entirely different forms with either green algae or cyanobacteria.

Pectenia atlantica Blue Felt Lichen

HOW TO SPOT Look for neat, large, scalloped slate-grey rosettes with darker centres on moss-covered trees and rocks.

DESCRIPTION **Foliose.** Thallus light lead-grey (dry) to dark grey (wet), up to 15cm, with distinctive undulating concentric waves like a tiny sea swell radiating from the centre of the thallus; lobes up to 3.5mm wide. The lower surface has a furry, pale to blue-black cushion of richly branched felt (the hypothallus, arrow in detail image), visible at the growing tips of the lobes. The upper surface has faint paler radiating lines and straight lobe edges which can often be followed towards the centre of the thallus *Reproduction:* **Isidia** which are warty, irregular and knoblike, but not flattened, densely covering the centre of the thallus in mature individuals.

WHERE On well-lit mossy trees in undisturbed oceanic woodland, occasionally on rocks in more exposed situations in the west of Scotland and western Ireland.

NOTES Felt Lichens are highly sensitive to acid rain and have been lost in most of England and Wales.

SIMILAR SPECIES

P. cyanoloma is similar except has reddish-brown to almost black apothecia with no isidia.

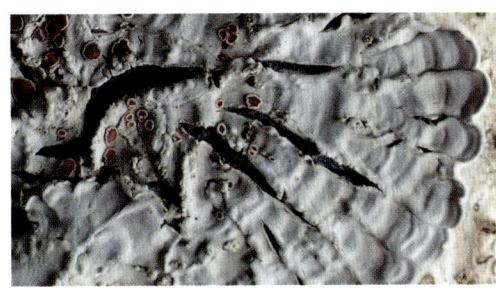

Foliose lichens 153

Pectenia plumbea Felt Lichen

HOW TO SPOT Small grey rosettes pressed closely to bark, with orange-brown apothecia centrally.

DESCRIPTION Foliose. Thallus light lead-grey (dry) to dark grey (wet). Similar to *P. atlantica* and *P. cyanoloma*, but usually smaller: thallus up to 5–10cm wide and lobes up to 2.5mm wide. Lobes have irregular networks of paler lines on the upper surface () and a felty blue-black cushion on the lower surface, often visible protruding from the edges (). *Reproduction:* **Apothecia** usually abundant, reddish to orange-brown, with a paler margin often visible (10×), up to 1mm wide.

WHERE On mossy broadleaved trees and rocks in moist, undisturbed woods.

NOTES This species is sensitive to pollution and declining throughout its range, now rare in England and Wales. In the east of Scotland, it is an old-growth woodland indicator.

SIMILAR SPECIES

P. atlantica and *P. cyanoloma* (p. 153) both have diagnostic concentric wave-like undulations on lobes, which are lacking in *P. plumbea*. *P. cyanoloma* has darker, almost blackish apothecia compared with *P. plumbea*. *Pannaria* species (p. 181) are similar in colour and habitat, but have smaller lobes and lack the thick felty cushion on the lower surfaces.

ATTACHED CENTRALLY; ONLY ON ROCK

Umbilicaria polyphylla Petalled Rock Tripe

HOW TO SPOT Chocolate brown clustered leafy lobes on upland silica-rich rock.

DESCRIPTION **Foliose.** Thallus chocolate brown, smooth, 2–5cm often forming larger patches, with lobes rising from centres and downturned at edges; lower surface black, minutely roughened. *Reproduction:* **Fragmentation** by black, loosely attached fungal structures specialised for asexual or clonal reproduction (🌀) produced on the lower surface.

WHERE On silica-rich rocks in the uplands and mountains, often found with Yellow Map Lichen.

NOTES Like other members of the genus *Umbilicaria*, lobes are attached centrally – thus the umbilicus in the name. This can be easier to see in other species that only form single lobes.

SIMILAR SPECIES

U. polyrrhiza (**A**) is also chocolate brown, smooth above and many lobed, but its lobes are thickly edged with stout, forked rhizines, protruding from below, appearing like a black fringe.

U. torrefacta (**B**) is brown, single-lobed and with an intricate pattern like puzzle pieces (🌀), and sometimes with lace-like intricate edges, up to about 4cm wide and black apothecia.

Foliose lichens

Umbilicaria cylindrica Fringed Rock Tripe

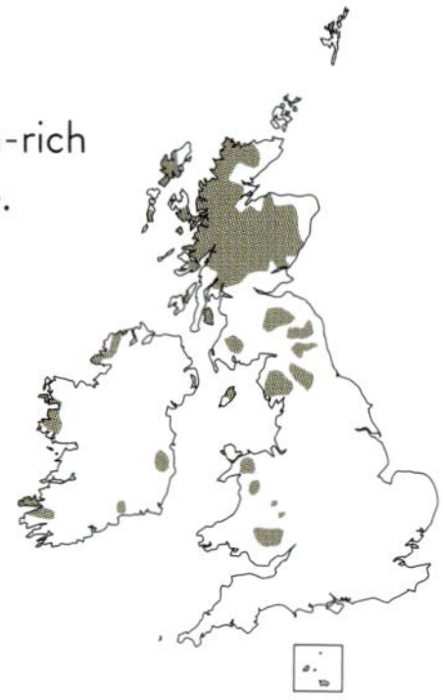

HOW TO SPOT Look for small circular patches of dove grey on silica-rich rocks in the mountains – the edges lift and usually have a bold fringe.

DESCRIPTION Foliose. Thallus dove grey, to brownish grey, up to 4cm wide and attached at a single, central point, with upturned edges fringed by stiff, branched, darker spine-like hairs; thalli are sometimes a single rounded lobe, sometimes several grouped lobes; lobes are heavily and finely textured with pale dusting of icing-sugar-like crystals (pruina, ). *Reproduction:* **Apothecia** are black, numerous, raised and distorted centrally like coiled liquorice ( ).

WHERE On silica-rich rocks in mountains and uplands.

NOTES The name *Umbilicaria* refers to the central attachment on the lower surface, the umbilicus in its name. When moist, a gentle turning tug lets you rotate the whole thallus a little bit around this point.

SIMILAR SPECIES

The other common greyish, pruinose Rock Tripe is *U. proboscidea* (Netted Rock Tripe), usually a single, rounded lobe, which lacks the fringe of branched hairs and has a central, raised area with net-like ridges ( ); also occurs in uplands and mountains on silica-rich rocks.

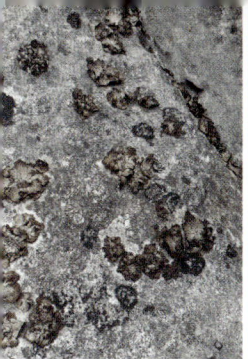

Dermatocarpon miniatum Common Stippleback

HOW TO SPOT Look for irregular groups of leathery grey lobes in seepage tracks on rock.

DESCRIPTION Foliose. Thallus grey (greener to brownish green when wet), with crowded and disorganised thick lobes, or rarely single rounded plates up to 5cm wide; arrangement of lobes usually very irregular. The thickish pale grey lobes have irregular edges and are covered with a fine powdery crystalline layer (pruina), and speckled with tiny black dots; the lower surfaces are pale brown and are connected to the rock only centrally. *Reproduction:* **Perithecia** are sunken in the lobes, and visible as the tiny black dots on the surface. The dots are the openings of these sexual, spore-forming structures.

WHERE On vertical or near-vertical lime-rich or only weakly acidic rocks in crevices or places often affected by water flows.

NOTES *Dermatocarpon* species are found on rocks which are moist to occasionally inundated, often beside streams and rivers. This is the largest and most common *Dermatocarpon* in Britain and Ireland; others are more specialist, rarer and harder to find. The other similar genus is *Umbilicaria* (pp. 155–56), found on silica-rich rocks in more exposed and drier habitats.

Foliose lichens 157

LOBES MEDIUM, UP TO 1cm WIDE

BD7

Nephroma laevigatum Mustard Kidney Lichen

HOW TO SPOT In western woods, not many brown lichens are relatively closely pressed to bark; this one has no obvious signs of reproduction when viewed from above.

DESCRIPTION Foliose. Thallus brownish-grey to milk-chocolate brown up to 8cm across and closely pressed to the bark near the edges, but with central mature lobes lifting. Lower surfaces are creamy brown and mostly smooth, without rhizines, hairs or veins (). *Reproduction:* **Apothecia** on the lower surfaces of mature central lobes, kidney-shaped and reddish brown.

WHERE On bark and mosses in old woodlands and on sheltered mossy rocks in coastal areas.

NOTES A cyanobacterial lichen and part of the Lobarion community, typical of old, moist, sheltered woods (p. 76). Mustard in the common name refers to the medulla, the inner cottony layer of the thallus, which is yellow in this species.

SIMILAR SPECIES

The similar **N. parile** differs only in having soredia on lobe edges (), sometimes in the centres of lobes as well. *Peltigera collina* (p. 149) is a similar size and colour palette with grey to brown tones, but has soredia on lobe edges and cottony-brown veins on the lower surface with tufted rhizines (). Always check the lower surfaces of brown or grey lichens in mossy woods!

The pathways

Melanelixia subaurifera Abraded Camouflage Lichen

HOW TO SPOT That's the tricky part! Usually very well hidden on bark, lying flat and often similarly coloured unless wet.

DESCRIPTION Foliose. Thallus olive green, olive brown or greenish, greener when wet, 3–5cm, with lobes closely attached to bark, forming rounded rosettes, with rounded tips. The surface of this species is slightly silky and bronzy looking, and although somewhat reflective, it is rarely what anyone would call shiny (more so in exposed situations). *Reproduction:* **Poorly formed isidia form in clumps within pale dot-like patches (soralia), which often become abraded, leaving pale, dot-like patches.**

WHERE On twigs and branches of broadleaved trees and on benches, railings etc., especially on horizontal surfaces.

NOTES One of the brownish to olive 'Melanelia' species, also called Camouflage Lichens (p. 71) because they are often tricky to spot. They can be learnt individually by their textures, degree of attachment, and reproductive structures. Get in the habit of using your lens to tell Camouflage Lichens apart.

SIMILAR SPECIES

M. glabratula (p. 133) usually grows on vertical bark, is often slightly shinier at lobe tips, and has isidia arising singly, not in abraded clumps. *M. fuliginosa* (p. 160) is usually on rock, darker brown and glossier with dense isidia. ***Melanohalea exasperata*** (right) has apothecia and well-spaced volcano-like isidia, and is tightly attached on twigs and branches. *Melanohalea laciniatula* (p. 71) has thin-textured lobes with isidia that flatten into miniature lobes as they develop.

LOBES MEDIUM, UP TO 1cm WIDE

BD4

Foliose lichens

Melanelixia fuliginosa Shiny Camouflage Lichen

HOW TO SPOT Dark brown rosettes on silica-rich rock stand out against the usual grey hues.

DESCRIPTION Foliose. Thallus dark brown to almost black, up to 3–5cm, with lobes closely attached to rock, forming rounded rosettes with rounded lobe tips. The surface of this species is glossy and reflective (), especially near lobe tips. ***Reproduction:* Peg-like isidia** are fine, cylindrical and form individually, becoming dense centrally.

WHERE On silica-rich rock, roofing tiles, memorials and sometimes worked timber.

NOTES Check the shape and development of reproductive structures to tell the Camouflage Lichens apart. If you can't tell them apart, just call them 'Melanelia' until you learn them! Their sunscreening pigment is melanin, and they darken in high light levels, just like human skin.

SIMILAR SPECIES

On rocks, look for dark forms of *Parmelia omphalodes* (p. 162), which is not so tightly pressed to the rock and has no isidia at all, but does have a network of pale lines and cracks (pseudocyphellae) near lobe tips (). Especially in the Cairngorm Mountains in Scotland, a very dark form of *P. omphalodes* is common, with a less distinct network of white lines and cracks (pseudocyphellae).

LOBES MEDIUM, UP TO 1cm WIDE

Xanthoparmelia verruculifera Warty Camouflage Lichen

HOW TO SPOT Large but can be difficult to spot! Its dull brown tones often mean it blends into the background of the rocks it grows on. Look for the leafy lobes covered in irregularly swollen knob-like isidia.

DESCRIPTION Foliose. Thallus medium to dark brown with grey or greener tones especially at growing tips, growing closely pressed to rocks, forming rosettes and wide-spreading patches up to 15cm, with lobes to 5mm wide. *Reproduction:* **Isidia** are coarse and knob-like arising from the surface of lobes (), sometimes becoming branched and opening at tips.

WHERE On coastal and inland silica-rich rocks and other well-lit and dry sites including memorials, roofing tiles etc.

BD 4

SIMILAR SPECIES

There are a few species of *Xanthoparmelia* on coastal rocks that differ in colour and reproduction. *X. loxodes* is very similar but almost exclusively coastal, usually paler yellow-brown, often with transverse (set at right angles) wrinkles on the lobes (), and isidia mounded in cauliflower-like outgrowths (). See the habitat image above which shows both species together; *X. loxodes* is on the left.

Foliose lichens

Parmelia omphalodes Smoky Crottle

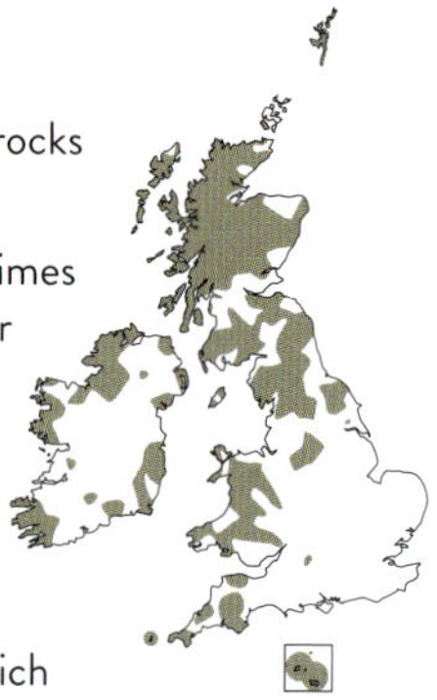

HOW TO SPOT Look for extensive blackish patches atop silica-rich rocks in the uplands.

DESCRIPTION Foliose. Thallus brownish grey to dark brown, sometimes blackish, or paler grey and greener in shaded situations, up to 12cm or merging into larger patches. Lobes up to 5mm wide, often squared at lobe tips, with a distinct network of white lines and cracks (pseudocyphellae,) usually more visible at growing tips; lower surfaces black with simple or forked black rhizines ().
Reproduction: Apothecia are sometimes produced, deep shiny brown and bowl-like with a margin the colour of the thallus. Black pores, which are the openings of asexual spore-bearing structures (pycnidia), can be abundant on the surfaces of lobes ().

WHERE On silica-rich rocks in the uplands.

SIMILAR SPECIES

P. saxatilis (opposite) is typically mostly grey with brown on lobe tips only, but has isidia arising from pseudocyphellae. Occasionally, *Platismatia glauca* (left) can be dwarfed and fully brown when in full sun, on mountain rocks or fenceposts, but its lobe edges are raised and curled.

Parmelia saxatilis Salted Shield

HOW TO SPOT Look for grey rosettes with darker centres on rocks, or extensive grey patches on tops of old stone walls, monuments or boulders.

DESCRIPTION **Foliose**. Thallus pale grey to bluish- or greenish-grey, often darker towards the centre, up to 12cm or thalli merging into larger patches. Lobes up to 5mm wide, often squared at tips, with a distinct network of white lines and cracks (pseudocyphellae, ●) usually more visible at growing tips, and with simple or forked black rhizines on the lower surface (●). Tips of lobes often tinged brownish. *Reproduction:* Isidia have darkened tips (●) and arise from pseudocyphellae or lobe surfaces. Older central areas can be covered with darker isidia, which sometimes obscure the lobes entirely; occasionally with brown, bowl-like apothecia up to 1cm diameter, with raised margins coloured the same as the thallus.

WHERE On acid bark, silica-rich rocks, monuments, walls and old wooden fences, and often abundant.

SIMILAR SPECIES

P. sulcata (p. 165) is similar, but has powdery soredia arising from pseudocyphellae, and lobes tend to rise from their substrate, which is more often tree bark. *P. omphalodes* (p. 162) can be grey to greenish-grey in the shade but has neither soredia nor isidia. *Xanthoparmelia conspersa* (right and p. 164) has isidia and grows on rock, but is yellowish green.

LOBES MEDIUM, UP TO 1cm WIDE

Xanthoparmelia conspersa Peppered Rock Shield

HOW TO SPOT Yellowish-green rosettes tightly pressed onto silica-rich rock stand out against grey and brown lichens.

DESCRIPTION Foliose. Thallus yellowish green, rosette-forming, and often coalescing into large patches; lobes square-tipped and flat, closely pressed to rock, up to 3mm wide, with a dark brown to black lower surface. *Reproduction:* Peg-like isidia are cylindrical to slightly swollen and sometimes branched (); they can be sparse to very dense, sometimes covering the interior of the thallus; frequently with chestnut brown, bowl-like apothecia in the centre.

WHERE On silica-rich rocks throughout Britain and Ireland.

NOTES The Rock Shield Lichens are hugely diverse worldwide, with dozens of yellowish-green species like this one growing on rocks, but some are brownish colours (see *X. verruculifera*, p. 161). Only two yellowish-green species are common in Britain and Ireland.

SIMILAR SPECIES

X. mougeotii is very similar, also only on rock, but much smaller and differs in its mounded clusters of soredia (soralia), produced on tops of lobes. Common in southern Britain.

The pathways

Parmelia sulcata Hammered Shield

HOW TO SPOT Probably one of our commonest lichens on bark; look for loosely attached greyish lobes, slightly squared at the tips. The lobes really do look like hammered metal!

DESCRIPTION Foliose. Thallus bluish grey to greenish grey, up to 8cm, sometimes in neat rosettes. Lobes up to 5mm wide, slightly squared and with irregularly distributed white lines and cracks near the tips (pseudocyphellae,), and black on the lower surface with black rhizines. **Reproduction: Powdery soredia** produced in clumps (soralia) from pseudocyphellae, forming round to irregular patches. Apothecia, rarely present, forming deep brown bowls with margins the colour of the thallus, up to about 6mm.

WHERE On broadleaved trees and shrubs, railings, and rarely on rocks; slightly more tolerant of nitrogen pollution than *P. saxatilis* (p. 163), so more abundant generally.

NOTES A thin coating of crystalline powder (pruina) is sometimes present on lobe tips, but tends not to be an important feature in *Parmelia*.

SIMILAR SPECIES

Punctelia (**A** and p. 166) and *Parmelia* (**B**) are both common rosette-formers on trees. Look for the different patterns on the lobe tips formed by their pseudocyphellae. If there are none, check *Hypotrachyna*, *Parmotrema*, *Parmelina* and *Hypogymnia*.

Foliose lichens

Punctelia subrudecta Speckled Shield

HOW TO SPOT You can sometimes pick this out at a distance among other large Shield Lichens by its slightly minty greenish colour and neat, round-edged rosettes.

DESCRIPTION **Foliose.** Thallus greenish-grey, lower surface pale to light brown, up to 5–10cm diameter, with smooth lobes to 8mm wide, with rounded tips and tiny pale flat dots on the lobe ends (pseudocyphellae,). *Reproduction:* **Powdery soredia** develop in tiny dot-like mounds (soralia), developing from pseudocyphellae on surfaces of lobes.

WHERE On bark of deciduous trees, and on wood fences and old benches, rarely on moss and silica-rich rocks; more common in the south and west, often in well-lit situations.

NOTES Look carefully at the pale dots on the lobe tips – these so-called punctuations give the genus its name. They may look a bit like grains of salt sprinkled on the young lobes.

SIMILAR SPECIES

Look out for the common *P. jeckeri* (**A**) with fewer pseudocyphellae and powdery soredia especially arising and developing on raised edges of lobes, and with very faint crystal dusting (pruina) at lobe tips ( ). *Cetrelia olivetorum* (**B**), with a strongly western distribution, has broader lobes with fainter dot-like pseudocyphellae and powdery crescents of soredia ( ) on upright ruffled lobe edges.

LOBES MEDIUM, UP TO 1cm WIDE

Flavoparmelia caperata Common Greenshield

HOW TO SPOT Saucer-sized apple-green or yellowish-green rosettes on trees stand out against the typically blue-grey colours of other lichens, usually visible even from high speed in a car.

DESCRIPTION Foliose. Thallus pale yellowish green, forming wide-spreading rosettes up to 20cm, with wide, rounded lobe tips and rumpled inner parts, often with fine wrinkles along with undulating lobes ( ); lobes sometimes exceeding 1cm wide. *Reproduction:* **Soredia** form from blistered, raised areas of the lobes that break open, so forming tiny rings of soredia ( ), described as looking a bit like scraped knuckles.

WHERE On tree trunks, branches, railings and even rocks, very abundant the further west you go.

NOTES The characteristic colour of this species is shared with *Usnea* (p. 68) along with other lichens, due to the presence of usnic acid, a sunscreening chemical in the upper cortex.

SIMILAR SPECIES

Look out for *F. soredians*, another abundant greenshield lichen especially in the far south, with smaller thalli, smaller lobes, and finer soredia forming lumpy and dense rounded clumps (soralia); it is chiefly southern but appears to be spreading northwards.

Foliose lichens

Parmelina pastillifera Button Shield

HOW TO SPOT A slightly smaller, neater bluish-grey Shield Lichen, with smooth lobe tips and with blackish tinges centrally, on parkland and wayside trees.

DESCRIPTION Foliose. Thallus pale bluish grey, sometimes almost whitish, 4–8cm, mostly neatly pressed against the bark, with lobes appearing smooth, distinctly wider at tips, rounded and scalloped, with small, neat gaps between lobes like the Loop Lichens (p. 71). *Reproduction:* Isidia which look like tiny black buttons, wider at their flattened tops ( ), are produced centrally on the surface of lobes.

WHERE On less acidic to nutrient-enriched and well-lit bark (e.g. ash, sycamore, field maple) or stones. Not widespread, but locally common.

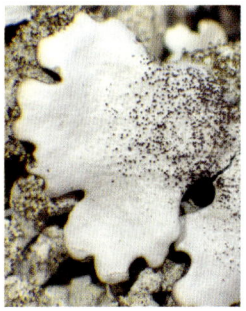

SIMILAR SPECIES

P. tiliacea is different in having more typical cylindrical isidia, brownish or grey brown, and in being slightly more loosely attached to the bark.

Hypotrachyna afrorevoluta Blistered Loop Lichen

HOW TO SPOT Smallish, blue-grey rosettes with scalloped outer lobes.

DESCRIPTION Foliose. Thallus bluish grey to greenish grey, up to about 5cm wide, with looping lobe edges especially when young, leaving neat, dark round spaces between the lobes, and tips often turned downwards (); lower surfaces black with black rhizines.
Reproduction: **Coarsely knobbly soredia** form on surfaces of raised lobes in the centre of the thallus, where blisters of thallus break open, like torn skin on bent knuckles ().

WHERE On branches of deciduous trees, sometimes forming individual rosettes, and sometimes covering large areas where many young thalli grow together. Sometimes on silica-rich rock.

NOTES This species along with *H. revoluta* (below) is often present along with other Shield Lichens on bark or rock, and often goes unnoticed. Look closely to find a smooth lobed (no white dots or cracks) rosette that would easily fit in the palm of your hand, and you might just spot it. See p. 71 for a few notes about Loop Lichens.

SIMILAR SPECIES

H. revoluta is similar and grows in similar habitats, even with *H. afrorevoluta*, but has lobes that rise up prominently, turn downwards and become covered in an even dusting of fine powdery soredia, so fine they are hard to distinguish even with a hand lens.

Foliose lichens

Hypotrachyna laevigata Smooth Loop Lichen

HOW TO SPOT Pale bluish-grey leafy lichens on bark particularly in the west, noticeable especially for the conspicuous rounded black gaps between the squarish thallus lobes.

DESCRIPTION Foliose. Thallus pale bluish grey to greenish grey, up to 15cm across, surface even and smooth with large lobes up to about 5mm wide, noticeably squared-off and more or less parallel sided; black branched rhizines can be thick on the lower surfaces of lobes (🔍). *Reproduction:* **Powdery soredia** are produced in rounded balls on lobe tips, a bit like tiny pom-poms.

WHERE On well-lit and acid-barked deciduous trees (birch, alder, oak), particularly common in the west.

NOTES The Loop Lichens are especially noticeable for their conspicuous and neatly rounded and curved gaps or 'negative spaces', which stand out in contrast to the pale lobes of the lichen. This species frequently forms saucer-sized thalli in the western woods of Scotland.

SIMILAR SPECIES

Several other loop lichens are also found in similar habitats: *H. taylorensis* (**A**) forms large patches up to 50cm across on tree bases and rocks with distinctive drooping ringlet-like tubes made of curled-over lobes; *H. sinuosa* (**B**) is a much smaller species especially on twigs, with a yellowish tinge and erect globular masses of powdery soredia on the tips of lobes (🔍). *Parmotrema* species (opposite) have simple rhizines and rounded lobe tips.

Parmotrema perlatum Powdered Ruffle Lichen

HOW TO SPOT Look for ruffled, blue-grey lichens with large rounded lobes, forming loose patches and sometimes rosettes.

DESCRIPTION Foliose. Thallus bluish grey to greenish grey, to 15–20cm diameter, with loosely attached, smooth lobes to 8mm wide, and black simple or forked hairs arising directly from edges of lobes (cilia; arrows); lower surfaces have a wide dark brown zone without rhizines and are black centrally or on older parts.
Reproduction: Powdery, **fine soredia** are produced in crescent shapes (soralia) at the edges of lobes ().

WHERE On acid bark, rocks and sometimes turf, more common in the south and west, often in well-lit situations.

NOTES The Ruffle Lichens have smooth, large lifting lobes, making messier rosettes or patches than other smaller Shield Lichens; they are usually visible from a couple of metres away.

SIMILAR SPECIES

Cetrelia olivetorum (p. 166) in temperate rainforests has similar soralia, but with white dot-like patches (pseudocyphellae) (). *P. crinitum* (**A**) is similar but looks distinctly bristly with long black hairs (cilia) arising from the tips of isidia and edges of lobes (), nicknamed 'Desperate Dan'. It prefers mossy trees and rocks or humid coastal rocks. *P. reticulatum* (**B**), in the far west and south, has a fine network of white lines () on the upper surface.

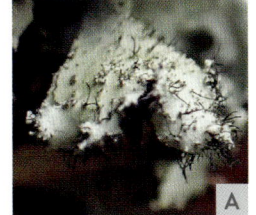

Platismatia glauca Frilly Lettuce Lichen

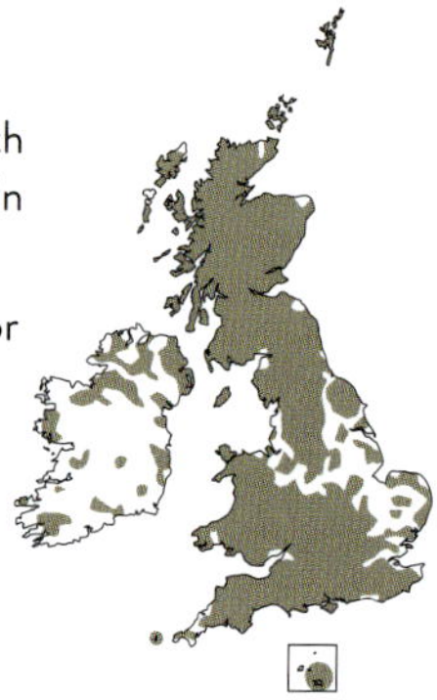

HOW TO SPOT Look out for large, lifting, paper-thin leafy lobes with crisped and divided lobe edges, a bit like curly kale, loosely attached in acid situations.

DESCRIPTION Foliose. Thallus pale greenish in shade to pale grey or sometimes brown in exposed situations, up to 15cm; lobes papery and thin, up to 1.5cm wide, rounded at tips, wrinkled, frequently crisped, crinkled and ruffled, rising at the edges. They are attached with a few rhizines only centrally which are hard to spot. *Reproduction:* **Isidia or soredia or both** on crisped edges of lobes.

WHERE On bark and rocks or wooden fences, sometimes on soil. Uniformly found in acidic or nutrient-poor situations and declining due to nitrogen pollution. Common on birch, conifers and alders.

NOTES This is an incredibly variable species, from all brown and small in full sun, to exuberant and lushly greenish in shade. Lower surfaces are equally varied, sometimes mostly white, often with large brown areas and sometimes black. Pinkish colours are frequent and caused by a parasitic fungus.

SIMILAR SPECIES

Particularly in the north, keep your eyes peeled for *Tuckermannopsis chlorophylla*, a smaller, uniformly brown lichen with a similar texture but smaller lobes and turning green when wet, and often tucked in among the larger lobes of *P. glauca*. The Ruffle Lichens – *Parmotrema* species – are similar in size and in the lifting nature of the lobes, but differ in texture, with thicker, duller, even-coloured lobes and never with white lower surfaces in Britain and Ireland.

Hypogymnia physodes Hooded Tube lichen

HOW TO SPOT Small and neat bluish-grey thalli, appearing to be outlined in black due to the black lower surfaces and inflated lobes.

DESCRIPTION Foliose. Thallus light bluish grey, whitish in very exposed situations, with a black and naked lower surface, up to about 5–6cm wide; lobes hollow, inflated, 2–3mm wide, flattening and widening and breaking open at the tips. *Reproduction:* **Soredia are coarse** (individually easily visible at 🔍) and produced on the insides of hollow lobe tips, as they open and curve upward and backward. These are described as lip-shaped soralia. Apothecia brown and bowl-shaped, found only occasionally.

WHERE On acid bark of trees, shrubs, heather, and on rocks and over heathy ground, only where nitrogen pollution is largely absent.

NOTES This species has been used for experiments to understand dispersal and establishment of soredia, possibly because of the large size of individual soredia grains. One researcher individually placed hundreds of soredia in different microsites to see how they stuck and grew – or didn't.

SIMILAR SPECIES

Hypogymnia tubulosa is similar but has narrower, linear lobes which remain tubular along their length (**A**), expanding to globular inflated tips, not breaking open, and covered in finer soredia (**B**, 🔍); it is often more frequent on twigs.

Foliose lichens

Menegazzia terebrata Port-hole Lichen

HOW TO SPOT Neat, even greenish-grey rosettes, with regular black holes in the upper surface on trees in the west.

DESCRIPTION Foliose. Thallus grey and small, up to about 8cm diameter, with a neat and even appearance, with lobes hollow, brown and rounded at the tips, and a black lower surface without rhizines. Neat black holes are present at the branch points of some more central lobes. *Reproduction:* **Powdery soredia** are produced in mounded or lip-shaped clusters (soralia) on the ends of lobes, arising as small projections on the lobes and becoming slightly protruding.

WHERE On acid bark of birch, alder and oak, rarely on conifers, sometimes on mossy rocks in temperate rainforests only in the west, especially Scotland.

NOTES There are only two species of *Menegazzia* in Britain and Ireland, both restricted to temperate rainforest habitats and both rather uncommon.

SIMILAR SPECIES

Hypogymnia physodes (p. 173) is much more common and also has hollow lobes with black lower surfaces, but lacks the neat port-holes in the upper surface.

LOBES NARROW, UP TO 2mm

Arctoparmelia incurva Sorediate Ring Lichen

HOW TO SPOT Bright but pale yellowish-green patches on upland silica-rich rocks, sometimes forming concentric arcs.

DESCRIPTION Foliose. Thallus yellowish grey to yellowish green, small up to 5cm, circular, with long, narrow contorted and tightly packed lobes that radiate outwards and curve under at the tips, sometimes overlapping each other. Lobes are convex, appearing rounded like fingers in surface view. Lower surfaces are pale. Larger patches can arise from thalli growing together, with centres falling away to leave arcs. *Reproduction:* **Powdery soredia** are formed in pale yellow raised and rounded mounds (soralia) found on tips of inner lobes, which are raised and downward curving.

WHERE On well-lit rocks and sometimes wood in upland regions and mountains. Increasingly found on human-made surfaces.

NOTES Locally abundant, but not widespread.

BD 4

SIMILAR SPECIES

Xanthoparmelia mougeotii (p. 164) is similar in colour and in being tightly pressed to rocks, but is not so restricted to the mountains. Its lobes are flatter with lower surfaces dark brown to black, and soralia are produced on the surfaces of lobes, not exclusively on tips. ***Parmeliopsis ambigua*** (right and p. 73) also has flattened, mostly non-overlapping lobes, but is mainly on acid bark or bare wood.

Foliose lichens

Anaptychia runcinata Fringe Lichen

HOW TO SPOT Chocolate-brown round patches on the upper shore on coastal rocks, mostly above the spray zone. Usually but not exclusively coastal, and characteristically pressed flat against the rocks.

DESCRIPTION Foliose. Thallus chocolate brown to dark brown (dry) or dull olive green (wet), forming rosettes up to 10–12cm across or extensive mats of radiating narrow lobes with a markedly dull or matt surface (). Lobes up to 2mm wide, appearing long due to their narrowness. *Reproduction:* **Apothecia** with dark discs and a raised margin the colour of the thallus, but not always present.

WHERE Mostly on exposed coastal rocks, well above the splash zone; less frequently inland and sometimes on trees in the west.

NOTES Usually visible from a couple of metres away as dark round patches, resolving into very narrow lobes only on close inspection.

SIMILAR SPECIES

A. mamillata (**A**) and *A. ciliaris* (**B**) both have a different look, with open tufts of strap-shaped lobes loose and rising from the surface they are growing on, and with long hairs (cilia) arising from the edges of lobes. *A. ciliaris* is a scarce and threatened species found inland on parkland trees in Dorset through to the Cotswolds. *A. mamillata* is a brownish coastal species on rocks, much rarer than *A. runcinata*.

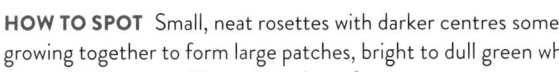

LOBES NARROW, UP TO 2mm

Physconia grisea Grey Frost Lichen

HOW TO SPOT Small, neat rosettes with darker centres sometimes growing together to form large patches, bright to dull green when wet, on city trees and human-made surfaces.

DESCRIPTION Foliose. Thallus whitish grey to dark or brownish grey (dry) to green (wet), 5–10cm, closely pressed to bark, with conspicuous frosting of powdery crystalline dusting (pruina) on lobe tips, best seen when damp or dry ( ); lobes flat, radiating, round-tipped, up to 2mm wide, matt, with simple white rhizines on the lower surface ( ). *Reproduction:* **Coarse and irregular soredia or fragile isidia** are formed on lobe edges, eventually covering centres of older thalli.

WHERE On nutrient-enriched bark and sometimes stone or human-made surfaces (window ledges, wall-tops, monuments) in well-lit situations.

SIMILAR SPECIES

The combination of crystal-dusted tips and green-when-wet, narrow-lobed rosettes is distinctive for the *Physconia* genus. ***P. distorta*** shown here has crystal-dusted apothecia and no soredia. Also look out for the smaller and usually browner *Phaeophyscia orbicularis* (p. 132) also turning green when wet but with oval patches of soredia and without pruina.

Foliose lichens

Physcia caesia Blue-grey Rosette Lichen

HOW TO SPOT This little bluish-grey lichen grows neat and flat, often on pavements and walls in human-made habitats.

DESCRIPTION Foliose. Thallus pale grey to bluish grey, up to 3cm, usually in neat rosettes, with lobes up to 1mm wide, matt, and slightly convex in cross-section, with tiny whitish, raised patches (pseudocyphellae,) easier to see when wet. *Reproduction:* **Powdery soredia** borne in mounded bluish clusters (soralia) on the surfaces of lobes; rarely with apothecia.

WHERE On pavements and mortar, marble gravestones and other nutrient-enriched places. Abundant and pollution tolerant.

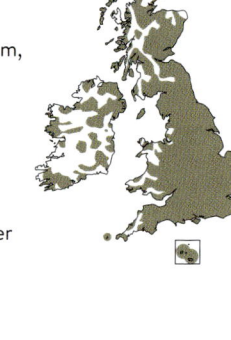

SIMILAR SPECIES

Watch out for ***Physcia dubia*** (Powder-tipped Rosette), which has its lip-shaped clusters of soredia on upturned lobe tips. *Physconia grisea* (p. 177) has soredia on the edges of thicker and wider lobes with crystalline-dusting (pruina,) on lobe tips.

Physcia aipolia Hoary Rosette Lichen

HOW TO SPOT Minute but neat blue-grey rosettes with blackish-brown discs on smooth bark.

DESCRIPTION Foliose. Thallus blue-grey, up to 3cm, with whitish irregular patches (pseudocyphellae) across the upper surfaces of lobes, more visible when wet (); lobes to 1mm, matt, convex in cross-section, pressed closely to bark. *Reproduction:* **Apothecia** dark brown to blackish, flat, with blue-grey slightly raised margins the colour of the thallus, up to 2.5mm, crowded centrally, sometimes with a thin white powdery crystalline covering (pruina,).

WHERE On twigs and branches with mildly acidic to neutral pH bark (e.g. ash, sycamore, cherry), rarely on old wood or walls.

SIMILAR SPECIES

Watch out for the much less common *P. stellaris*, which is similar but lacks the white patches when moist, and *P. leptalea* (p. 180), with loosely attached lobes and long cilia, both with apothecia and without powdery soredia. *Physconia distorta* (p. 177) has white crystal-dusting on discs and lobe tips and turns green when wet.

Physcia tenella Fringed Rosette Lichen

HOW TO SPOT Look for tiny blue-grey lichens on twigs, especially growing near the orangish-yellow *Xanthoria parietina*.

DESCRIPTION Foliose. Thallus pale bluish-grey, 2–3cm, of strap-shaped lobes, these with long pale or dark-tipped cilia arising from edges; lobes matt, with tiny whitish raised areas, more conspicuous when wet (pseudocyphellae, 🔍). *Reproduction:* **Powdery soredia** produced in lip-shaped clusters (soralia) at the tips of lobes; apothecia rare, but when present are dark brown with margins the same colour as the thallus.

WHERE On well-lit bark, twigs, wood, concrete, limestone etc., especially with nutrient enrichment.

NOTES This tiny blue-grey lichen has expanded markedly with declining sulphur dioxide and increasing nitrogen pollution, covering twigs across large areas, especially near farms, along roadways and in gardens and parks in cities.

SIMILAR SPECIES

P. adscendens (**A**) is similar but has soredia produced underneath lobe tips, which are inflated, hood-shaped and paler (🔍). It is more common on rocks than *P. tenella*, but both are often found together on nutrient-enriched bark and twigs.

P. leptalea (**B**), a common coastal species in the southwest, is similar, but lacks soredia entirely and has neat blackish apothecia.

The pathways

Pannaria rubiginosa Brown-eyed Shingle Lichen

HOW TO SPOT Look for small blue-grey rosettes on mossy bark in western woods, speckled with tiny reddish-brown discs.

DESCRIPTION Foliose. Thallus pale blue-grey (dry), bluish to slate-grey (wet), up to 2–4cm across, with narrow, radiating lobes to 2mm wide on thallus margins. A mostly flat black zone of radiating fungal hyphae is sometimes visible at the growing edge of the thallus ( ). *Reproduction:* **Apothecia** reddish brown, formed on older central parts can reach 1.5mm wide, with a distinct raised and crinkled margin the colour of the thallus ( ).

WHERE On mossy or bare bark of broadleaved trees in sheltered areas in the west, where moisture is plentiful.

NOTES This is a cyanobacterial lichen, so it is only found in sheltered, moist microhabitats.

SIMILAR SPECIES

P. conoplea is similar in habitat, size and colour, but differs in forming coarse, diffuse soredia ( ). It is an indicator of ancient woodland. *Pectenia* species (pp. 153–54) have larger lobes and have a thick (not flat) felty cushion of richly branched hyphae on the lower surface of lobes ( ).

Foliose lichens

Cladonia subcervicornis Two-tone Cladonia

HOW TO SPOT From a distance the mottled patches of erect squamules in crevices on rocks appear bluish green when moist, and leaden grey-and-white when dry, sometimes with a purplish hue.

DESCRIPTION Squamulose Thallus grey to blue-grey above, sometimes appearing ever so slightly purplish (dry) to distinctly bluish (moist/wet), in patches up to 8cm. Made up of dense, erect squamules, longer than wide, up to 0.5–2cm long. Lower surfaces have subtle greyish veining (🔍) and are blackened at the base with bright white tips, conspicuous when dry as they curl over.
Reproduction: **Apothecia** are found occasionally on short and often irregular erect stalks scarcely taller than the squamules; stalks have small cups and minute dark brown apothecia.

WHERE In pockets of thin soil on silica-rich rocks and mossy walls.

NOTES Forming even-topped cushions, often identifiable at a distance by the lead-grey colour, sometimes with inconspicuous stalks. From above, often two-toned, as both upper and lower surfaces are visible at the same time, each a different colour.

SIMILAR SPECIES

Other species of *Cladonia* form dense cushions of squamules, but no others have the characteristic combination of blue tinge above, cottony white lower surface and blackening at the base. *C. strepsilis* (not shown) forms dense convex cushions easily dislodged from soil, but has a yellowish-brown hue, with highly textured upper surfaces of squamules (🔍).

Hypocenomyce scalaris Common Clam Lichen

HOW TO SPOT Extensive patches of yellowish brown on conifer bark or old wood, especially near the bases of trees.

DESCRIPTION **Squamulose.** Thallus yellowish cream to hay-coloured or brownish yellow, forming irregular patches often over 10cm, made up of convex scales about 1mm wide, often tightly packed and slightly overlapping like roofing shingles, each rounded on the exposed edge. *Reproduction:* **Powdery soredia** fringe scale edges; apothecia greyish black, with contorted margins, sometimes abundant.

WHERE On bases of old conifers or on old wood, including fenceposts, fallen oaks etc. Usually on large and stable substrates. Sometimes on brick or stone.

SIMILAR SPECIES

Look out for similar but far less common *Xylopsora friesii*, composed of crowded, shiny convex islands of thallus (areoles), without soredia, but with contorted shiny apothecia, only in ancient Caledonian pinewoods.

Normandina pulchella Elf Ears

HOW TO SPOT Look on mossy trees or on the thalli of large foliose lichens in damp western woods or parklands for tiny, neat bluish scales that look like round little ears!

DESCRIPTION **Squamulose**. Thallus bluish green, bluer when dry, of tiny individual rounded or ear-shaped scales (squamules) with neat, raised pale lobe edges and faint, concentric growth rings ( ); individual scales up to 2mm. The colour is unique and distinctive.
Reproduction: **Powdery soredia** formed on larger scales, which sometimes conceal tiny perithecia, which project from the lower surfaces of the scales.

WHERE On mossy trunks and rocks in woodlands and parklands, growing over mosses, liverworts and other lichens.

NOTES The aptly chosen species name pulchella means little and beautiful. This species is increasing in its range, spreading eastwards and into urban areas.

SIMILAR SPECIES

The scale-like squamules of *Cladonia* (p. 66) can grow with this species, but they tend to rise from the surfaces they grow on, and they lack a neat, raised lobe edge.

CRUSTOSE LICHENS

We separate crustose lichens by a combination of colours, forms and reproductive structures, summarised in the Quick Guide of thumbnail images on the next page. As with the other growth forms, this guide will help you narrow down your observation to the best-fitting options. We first separate out the bold colours from more muted tones; choose the closest colour of your lichen as observed from a metre or so away. Within each colour group, we first present species with lobed margins, followed by sets of species that have similar reproductive structures. Find the thumbnail image that best matches your lichen and go to the page numbers cited to read the Species Descriptions and check that your specimen fits the other features described.

1. **Colour** As with the other growth forms, bold colours are typically separated out easily.

- **Orange** and **yellow** in our usage refers to *either or both the thallus (body of the lichen) or apothecia (fruiting bodies)* having these vivid colours.
- **Brown** and **black** colours refer to the *thallus*, not the fruiting bodies (apothecia or perithecia).
- **Red** or **reddish orange** colours of the thallus are not uncommon on silica-rich rocks, but many of these require microscopy for identification. We present a small selection on page 233.
- If your lichen is none of these bold colours, white and pale grey are grouped together along with greenish colours, as they can be difficult to assign to one or another and sometimes vary. Lichens that appear pale green for one person can be greyish for another. *Check forms and reproductive structures.*

2. **Form** Within each colour group, **lobed** species of crustose lichens are presented first; remember, even though their lobe tips may appear foliose, they cannot be peeled up at the edges with a fingernail because they are attached across their whole lower surface.

3. **Reproductive structures** Finally, we arrange species into sets according to shared reproductive structures. Recall that any structure that produces spores is referred to as a **fruiting body**.

- **Fruiting bodies: apothecia with open, round discs** In lichens, sexual reproduction most often produces **apothecia**, which are usually rounded with exposed discs.
- **Soredia or isidia** A large number of crustose lichens produce **soredia**; fewer crustose lichens produce **isidia**. All those that exhibit vegetative reproduction are grouped together.
- **Fruiting bodies: other** A few reproductive strategies are represented by only one or two species among crustose lichens in this guide:

    - **apothecia** in thallus warts (*Pertusaria*) so with hidden discs
    - **apothecia** stalked (*Baeomyces*)
    - **apothecia** elongate (*Graphis, Arthonia*)
    - **perithecia** (*Pyrenula, Verrucaria* and relatives), which are closed and usually black, sometimes sunken in the thallus, opening by a tiny pore
    - **pycnidia**, which are tiny and produce asexual spores and are the dominant reproductive mode of *Lecanactis*
    - **mushrooms** (*Lichenomphalia*)

QUICK GUIDE
CRUSTOSE LICHENS

THALLUS OR FRUITING BODIES YELLOW TO ORANGE

Variospora (pp. 188–90)   *Candelariella* (p. 191)   *Candelariella* (p. 193)   *Athallia* (p. 194)

Rhizocarpon (p. 192)   *Protoblastenia* (p. 195)   *Flavoplaca* (p. 196)   *Chrysothrix* (p. 197)

THALLUS BROWN TO BLACK

Acarospora (p. 229)   *Rhizocarpon* (p. 212)   *Verrucaria* (p. 230) and *Hydropunctaria* (p. 232)   *Placynthium* (p. 231)

THALLUS REDDISH

Rusty reds (p. 233)

THALLUS PALE – WHITISH TO GREEN OR GREY
FRUITING BODIES: OTHER

Pertusaria (p. 221)   *Baeomyces* (p. 222)   *Arthonia* (p. 223)

THALLUS PALE – WHITISH TO GREEN OR GREY

THALLUS LOBED

Protoparmeliopsis (p. 198)

Solenopsora (p. 199)

Diploicia (p. 200)

Kuettlingeria (p. 201)

Placopsis (p. 202)

FRUITING BODIES: APOTHECIA WITH OPEN ROUND DISCS

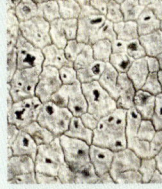
Circinaria (p. 203)

Tephromela (p. 204)

Ochrolechia (p. 205)

Lecanora (pp. 206–208)

Ophioparma (p. 209)

Lecidella (p. 210)

Mycoblastus (p. 211)

Rhizocarpon (p. 212)

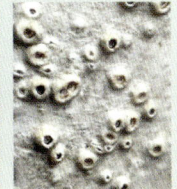
Thelotrema (p. 213)

Icmadophila (p. 214)

SOREDIA OR ISIDIA

Porpidia (p. 215)

Ochrolechia (p. 216)

Lepra (p. 217)

Lepraria (p. 218)

Lepra corallina (p. 219)

FRUITING BODIES: OTHER

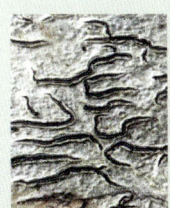
Graphis (p. 224)

Lecanactis (p. 226)

Pyrenula (p. 227)

Lichenomphalia (p. 228)

Variospora flavescens (Caloplaca flavescens) Limestone Firedot

HOW TO SPOT Bright yellowish-orange or orange patches on stone, often forming arcs, and usually with a whitish zone behind the growing tips and with darker orange apothecia.

DESCRIPTION Lobed Crustose. Yellowish-orange thallus with finger-like lobed tips, in neat rounded patches up to from a few cm to 10cm across, but sometimes growing together or as arcs after older central parts fall away. Lobes are narrow and often long, more than 2mm and convex, often whitening in a zone behind the growing tips. Older parts are broken into separate islands of thallus (areoles). *Reproduction:* **Apothecia** usually present and darker orange.

WHERE On limestone, walls, cement, mortar and other lime-rich rocks; probably the commonest lobed Firedot.

NOTES A number of Firedots are lobed at the tips, and the differences between them have to do with the shapes – lengths and cross-sections – of lobes, reproductive structures and the presence of a dusting of crystals on the surface. The lobed species presented here and on the facing page are all found on lime-rich rocks. Check lobe tips ( ) to tell them apart.

SIMILAR SPECIES

Variospora aurantia (**A**) has a yellower colour, with flattened, spreading lobe tips, as if they have been hammered. *Calogaya pusilla* (*Caloplaca saxicola*) (**B**) has a pinkish-orange colour, with shorter lobe tips and a dusting of fine crystals (pruina) on lobe tips. *Calogaya decipiens* (**C**) also has pruina but has soredia erupting from small inner lobe tips and eventually covering inner areoles. *Rusavskia elegans* (p. 140) is dark orange with well-separated lobe tips, which are actually foliose, so lobes can be peeled up.

Variospora thallincola (Caloplaca thallincola)
Coastal Firedot

HOW TO SPOT Bright orange, neat round rosettes growing completely attached on seaside rocks.

DESCRIPTION Crustose. Thallus bright orange, up to 3–4cm, sometimes growing together to create larger patches. Edges of thallus formed of neat, regular, convex finger-like lobes closely attached on rocks, with a centre of irregular tiny islands (areoles) often with blackish areas. *Reproduction:* **Apothecia** neat, up to 1mm, margin and disc all of the same rich bright orange as the thallus.

WHERE Only on coastal rocks, in the orange splash zone. Sometimes overgrowing the middle-shore black Tar Lichen *Hydropunctaria maura* (p. 232).

NOTES This is one of the lichens that make up the 'orange zone' on coastal rocks. These lichens have an orange pigment called parietin as a sunscreening compound, which can be detected with a chemical test.

SIMILAR SPECIES

Other orange splash zone species include ***Flavoplaca marina*** (**A**), which has the same colour and has abundant apothecia, but lacks distinct finger-like radiating lobes; ***Polycauliona verruculifera*** (**B**) has similar thalli with long, radiating closely attached lobes in neat rosettes or rings and also has apothecia, but is usually a mustard-yellow colour and has large, rounded isidia (). It often forms rings where older thallus centres fall away.

THALLUS OR FRUITING BODIES YELLOW TO ORANGE; LOBED

Candelariella medians Lobed Goldspeck

HOW TO SPOT Bright lemon yellow, small rosettes with radiating lobes closely pressed onto human-made surfaces, especially in villages and churchyards and on limestone.

DESCRIPTION Lobed Crustose. Thallus lemon yellow to becoming blackish centrally, up to 3cm, forming irregular rosettes, with radiating short lobes at the growing thallus edge; edges of lobes are pressed together, and have a minutely roughened texture due to a fine crystalline layer (pruina, 🔍); thallus broken into areoles centrally and sometimes developing blackened cracks between islands of rounded granules. *Reproduction:* **Isidia** are rounded and smooth, in the centre of the thallus; apothecia are yellow, with a yellow margin, but are only rarely found.

WHERE This species occurs in areas with relatively high nutrient input, and on lime-rich rocks such as limestone, marble tombstones and concrete.

SIMILAR SPECIES

The bright lemon-yellow colour of young thalli and younger parts of older thalli is distinctive and contrasts with orange colours of Firedots which are often found nearby. The yellowish *Polycauliona verruculifera* (opposite) may appear similar, but that species is strictly on maritime rocks in the splash zone. *Candelaria concolor* (p. 141) is the same colour, but with incredibly fine, and loose foliose lobes and almost always on bark.

Crustose lichens

Rhizocarpon geographicum Yellow Map Lichen

HOW TO SPOT Brilliant greenish-yellow and black, finely patterned crusts on silica-rich rocks, abundant in the uplands, often noticeable from a few metres away.

DESCRIPTION Crustose. Thallus bright yellow or bright greenish yellow, usually up to 2–3cm but forming larger mosaics of neat, flat, smooth tiny islands (areoles, 🔍) on a black background, with the whole thallus usually edged in black. Thalli growing together can look like miniature-scale maps with black and yellow territories and black boundaries between them. ***Reproduction:*** **Apothecia** black, up to 1.5mm across, round to angular in shape with black margins (🔍).

WHERE On silica-rich rocks, wall tops, roofing slates and boulders in sunny locations from seashores to mountaintops, and often abundant.

NOTES A very variable species. Many *Rhizocarpon* species have neatly patterned thalli of regular areoles on black backgrounds; they are particularly diverse in the mountains.

SIMILAR SPECIES

Compare **R. geographicum** (on the left) with **R. lecanorinum** (on the right), which has crescent-shaped areoles that partially and individually surround apothecia.

192 The pathways

Candelariella vitellina Common Goldspeck

HOW TO SPOT Patches of dull golden mustard yellow in nutrient-enriched places, sometimes extensive.

DESCRIPTION Crustose. Thallus yellow to brownish-yellow or orangish-yellow, of rounded and slightly flattened, sometimes minutely lobed areoles, so small as to be termed granules, individually smooth rather than powdery, individually visible only at 10×, but *en masse* often conspicuous. **Reproduction: Apothecia** often present, up to 1.5mm but often smaller, with a brighter yellow margin becoming increasingly contrasting as discs darken with age.

WHERE On nutrient-enriched rocks, fenceposts (bird-perches), wood, tarmac or other human-made surfaces.

NOTES A few other yellow *Candelariella* species occur in Britain and Ireland, most of which are similar mustardy-yellow colours. Several form granular (tiny, rounded and smooth –) areoles or crusts which are covered in soredia.

SIMILAR SPECIES

C. aurella, visible as a collection of tiny lemon-yellow apothecia ( ), has a very thin scattering of smooth but tiny granules on a black background and is common on concrete, bricks and mortar in towns.

Athallia holocarpa Firedot

HOW TO SPOT You probably will only see this when examining something else with your hand lens, as the largest thalli are only about 1cm across.

DESCRIPTION Crustose. Thallus usually not apparent or merely a faint yellowish stain. *Reproduction:* **Apothecia** closely spaced to crowded, tiny and orange, less than 1mm across. Young thalli will have widely separated apothecia with relatively thick margins, becoming crowded and angular as they grow together.

WHERE On a wide variety of surfaces including tree bark and silica-rich to lime-rich rock; common on almost any nutrient-rich habitat, even seashells or leather.

NOTES This is a collection of related species, each with distinct habitat preferences, all with tiny orange apothecia and no apparent thallus.

SIMILAR SPECIES

Blastenia crenularia has large dark grey thalli up to 10cm and larger rusty-orange or brownish-red apothecia and is found on silica-rich stone, anywhere from the seashore to far inland.

Protoblastenia rupestris   Rotten Oranges

HOW TO SPOT   Often visible as a pale stain on concrete or mortar or as orange dots on limestone.

DESCRIPTION   **Crustose**. Thallus dirty grey brown or grey green, cracked, sometimes indistinct. *Reproduction:* **Apothecia** are orange with a slightly convex surface, irregular and without distinct margins (); apothecia thin towards the edges, appearing a bit like melted wax and often discolour with age ().

WHERE   On lime-rich surfaces such as mortar, concrete, limestone, and chalk pebbles.

NOTES   Most orange apothecia belong to the Firedots ('*Caloplaca*' group), but *Protoblastenia* is not at all related to these. This genus can often be distinguished by the lack of obvious apothecia margins.

SIMILAR SPECIES

Compare the Firedots *Athallia holocarpa* and *Blastenia crenularia* (opposite), both with obvious apothecia margins ().

Crustose lichens   195

Flavoplaca citrina (Caloplaca citrina) Mealy Firedot

HOW TO SPOT Yellow to orange granular crust forming extensive patches on vertical walls of brick or concrete. Little structure becomes clear even on close inspection.

DESCRIPTION **Crustose**. Thallus yellow to orange with no clearly defined borders, so can extend over tiny to very large areas, with surface sometimes broken into minute islands (areoles, 🔍).
Reproduction: **Coarse soredia-like granules** make up most or all of the thallus, and darker-orange apothecia are sometimes located after careful inspection (🔍), up to 1mm.

WHERE On concrete, mortar, nutrient-enriched rocks or brick, especially on vertical surfaces. Abundant.

NOTES This name applies in its broad sense to a suite of closely related Firedots that have thalli mostly made up of soredia. These vary in colour from scrambled egg to rich, deep orange.

SIMILAR SPECIES

The unrelated bright greenish yellow *Psilolechia lucida* is usually found in shaded underhangs on silica-rich rock or monuments, avoiding direct rainfall. See also Common Goldspeck *Candelariella vitellina* (p. 193), which is mustard yellow but with slightly flattened and expanded granules (🔍).

Chrysothrix candelaris Gold Dust

HOW TO SPOT One of the few '60mph lichens' that you can spot from a fast-moving vehicle. Brilliant yellow patches on the dry sides of rough-barked broadleaved trees can only be this species.

DESCRIPTION **Crustose**. A lichen made up of only soredia (leprose growth form) of a distinctive bright-yellow colour. Under a 10× lens, soredia may be diffuse or clumped, and are very often mixed with other leprose lichens of other colours. ***Reproduction: Soredia**.*

WHERE On rough bark, under large boughs or on the protected side of old broadleaved trees, particularly in bark crevices, but spreading more widely where sheltered from rain.

NOTES A handful of distinctive lichens occur in the same dry-side-of-old-trees habitat, avoiding direct wetting from rain, and it is worth looking out for them. *Lecanactis abietina* (p. 226) forms extensive pinkish-grey sheets. In the south, look for the black apothecia of *Cresponea premnea* (p. 210).

TOP TIP

Look for tiny yet distinctive pin lichens ( ), which might be found tucked deep in crevices of bark, often with *Chrysothrix candelaris*. The commonest one is ***Calicium viride*** with stalked apothecia up to 2mm tall that look like tiny black nails or pins arising from greenish-yellow areoles.

wet

Protoparmeliopsis muralis (Lecanora muralis)
Chewing Gum Lichen

HOW TO SPOT Pale patches on paving stones, pavements and kerbstones in cities.

DESCRIPTION Lobed Crustose. Thallus pale grey to yellowish green, up to 8cm, often smaller, with radiating flat lobes at the edges – you may not notice the lobes until you check closely (🔍).
Reproduction: Apothecia pale greenish to brownish, abundant centrally, with margins the same colour as the thallus (🔍).

WHERE On nutrient-enriched and lime-rich rocks, walls and other human-made surfaces, and sometimes sawn wood.

NOTES Think you've seen some chewing gum on the pavement? Look again to see if it's this canny lichen. There are a handful of other lichens common on pavements, especially *Lecanora campestris* (p. 207, **B**), a Rim Lichen with a faint white edge (🔍) and deep brown apothecia, and *Circinaria contorta* (p. 203), which forms clustered, but separate, tiny whitish spots each with a darker spot, the disc of the sunken apothecia.

Solenopsora candicans Limestone Solenopsora

HOW TO SPOT Look for small bright white rosettes with black dots on limestone rock.

DESCRIPTION Lobed Crustose. Thallus chalk white, up to 5cm, cracked into tiny islands (areoles) centrally but with radiating flat lobes, slightly widened at tips (👁), but remaining totally attached to the rock underneath. Completely and thickly covered in fine white crystals (pruina, 👁), so appearing chalky-textured. ***Reproduction:*** **Apothecia** black, but often appearing greyish or purplish-grey because of a pale crystalline covering (pruina), with whitish margins when young (👁). This layer may disappear as apothecia age.

WHERE On hard, sunny limestone and monuments.

NOTES Often found with orange Firedots (p. 80) and black *Verrucaria nigrescens* (p. 230).

SIMILAR SPECIES

Diploicia canescens (p. 200) is often just slightly bluer, with lobes appearing pleated, rather than spreading at the tips; it has soredia and only rarely forms apothecia, but these are black, with black margins.

Crustose lichens

Diploicia canescens White Pleated Lichen

HOW TO SPOT Conspicuous, white, neat rosettes, tight against rocks – often forming arcs where the older centres fall away.

DESCRIPTION Lobed Crustose. Thallus pale grey, whitish or slightly bluish-grey (dry), or with a greenish tinge (wet), of neat, radiating rosettes up to 5–6cm but often coalescing, matt textured. Edges of thallus have convex, long narrow lobes, slightly widened at the tips (🔍); centre of thallus irregularly lumpy (🔍). ***Reproduction:* Coarse powdery soredia** are in clusters (soralia) arising from tips or edges of inner lobes (🔍). Black apothecia with black margins sometimes found when thalli growing near the coast.

WHERE On lime-rich rocks like sandstone and limestone, and sometimes on old tree bases.

NOTES Probably the commonest of a very few whitish crusts with lobed edges; the smooth even colour, convex lobes and soredia arising from lobe edges and tips should be enough to distinguish it. Sometimes overgrown with free-living green algae where nutrient enrichment is high.

SIMILAR SPECIES

On rock, other whitish lobed crusts include *Solenopsora candicans* (p. 199) with blackish apothecia and without soredia, or *Kuettlingeria teicholyta* (opposite) with flatter lobes, darker centres and dispersed soredia (not clearly clustered).

Kuettlingeria teicholyta (Caloplaca teicholyta) Cryptic Firedot

HOW TO SPOT Look for small grey rosettes with white edges on concrete and other lime-rich surfaces.

DESCRIPTION Lobed Crustose. Thallus grey centrally, paler or whitish on edges, up to 5cm, cracked into tiny islands (areoles) centrally but with radiating flat lobes, slightly widened at tips (), and remaining completely attached to rock underneath. *Reproduction:* **Obscure and dispersed soredia** scattered across central eroded parts of thallus (); apothecia orange-brown to reddish-brown, rare but conspicuous when present.

WHERE On lime-rich stone, mortar, buildings, walls and monuments.

NOTES Most Firedots have conspicuous orange or yellow thalli or apothecia visible most of the time; this species is an exception in only producing such pigments in apothecia, and these being rare.

SIMILAR SPECIES

Solenopsora candicans (p. 199) has black apothecia; *Diploicia canescens* (opposite) has convex thallus parts and clustered soredia.

Placopsis lambii Pink Bull's Eye

HOW TO SPOT Small, neat radiating pinkish-grey crusts with irregularly lobed, central thickenings in the thalli surface – distinctive on upland silica-rich rocks.

DESCRIPTION Lobed Crustose. Thallus pale grey, 1–5cm in diameter, with radiating lobes on the thallus edge and cracked areoles (tiny islands) centrally (); raised, irregularly lobed pinkish to reddish cyanobacteria-containing structures (cephalodia; **1)** form in the centre of the thallus (). ***Reproduction:* Apothecia or soredia.** Apothecia (**2**) form a ring halfway between margins and cephalodia, with pinkish-orange discs and thick margins the colour of the thalli (); darker powdery grey to greenish soredia form in crater-like patches (soralia; **3**) irregularly distributed across the surfaces of lobes ().

WHERE On metal-rich silica-rich rocks in the uplands, and often in high-rainfall areas.

NOTES Only relatively few groups of lichens can form stable associations with cyanobacteria and green algae; cyanobacteria offer an in-house source of usable nitrogen in nutrient-poor habitats, while the green algae throughout the rest of the thallus provide sugars from photosynthesis.

SIMILAR SPECIES

Other pale grey radiating crusts are almost exclusively on lime-rich rocks.

THALLUS PALE – WHITISH OR GREY TO GREEN; APOTHECIA ROUND

Circinaria calcarea (*Aspicilia calcarea*)
Calcareous Rimmed Lichen

HOW TO SPOT One of the commonest large white crusts on limestone and marble, including tombstones. White or pale grey crustose patches can be seen from metres away.

DESCRIPTION Crustose. Thallus white or pale grey, large, up to 40cm, with smooth surface divided into little islands (areoles, ). Edges of thallus have a radiating fringe with a darker grey edge, and cracks between areoles (tiny islands in thallus) extend all the way to thallus edges ( ). *Reproduction:* **Apothecia sunken**, with white margins and black discs, often with a dusting of powdery crystals (pruina) on the disc surface and often more than one apothecium per areole ( ). Young apothecia have raised margins and concealed discs, with discs becoming irregular and exposed at maturity, and raised margins becoming less clearly visible.

WHERE On lime-rich (calcareous) rocks including hard limestone, walls and tombstones, but not found on cement or mortar.

NOTES Sometimes forms mosaics with conspicuous dark lines between thalli. Often encountered with orange Firedots (p. 80) and black *Verrucaria nigrescens* (p. 230) on pale-coloured rocks. Part of the larger group of sunken-disc lichens, once all called *Aspicilia*.

SIMILAR SPECIES

C. contorta also has sunken discs, each in a well-separated areole, and is found on a wider range of lime-rich substrates, including paving slabs, pavements and sandstones.

Crustose lichens 203

Tephromela atra Black Eye Lichen

HOW TO SPOT Look for black-and-white crusts on rocks – this one has coal-black, white-rimmed apothecia on a thick white thallus.

DESCRIPTION Crustose. Thallus white or pale grey, often lumpy like thick porridge, forming patches up to 8cm, sometimes with a dark border, with convex tiny islands of thallus (areoles,). *Reproduction:* **Apothecia** numerous, usually crowded centrally and becoming irregular in outline, up to 3mm, with coal-black discs and whitish margins matching the whitish thallus ( ). The internal tissues of the apothecia are purple-black from the upper layers right to the base.

WHERE On silica-rich rocks. Common throughout, especially on the seashore.

NOTES This species was originally named within *Lecanora*, but in fact belongs to a different family entirely. The species name 'atra' means black.

SIMILAR SPECIES

Lecanora gangaleoides is a lookalike species, but its apothecia tend to be smaller and rounder. To tell them apart, slice an apothecium in half with your fingernail or penknife and look for the colours inside ( ): purple-black all the way from top to base of disc is *Tephromela atra*, and a thin greenish black layer only on the top with white below is *L. gangaleoides*. Look out for the pale yellowish green parasitic lichen *Lecanora sulphurea* (p. 208) which often starts life on *T. atra* and later becomes independent.

Ochrolechia parella Crab's Eye

HOW TO SPOT Large creamy white thalli with abundant large thick-rimmed apothecia looking like white 'O's, on rocks.

DESCRIPTION Crustose. Thallus greyish to creamy white, large, thick, smooth to warted, sometimes covering large areas, especially on gravestones near the coast. With a distinct zoned edge of concentric bands of slightly different colours, edged in white. *Reproduction:* **Apothecia** abundant, up to 2–5mm, with a prominently raised thick margin, disc often the same colour as the thallus due to thickly encrusting crystalline coating (pruina, 🔍), or pale pinkish brown within a paler margin. Apothecia can be well spaced or crowded and angular.

WHERE Coastal rocks are the main habitat of this species, but it is abundant throughout on walls, brick, slate and occasionally even on tree trunks.

NOTES Although the name *Ochrolechia* has to do with the creamy colour of the discs, not their shape, you can always remember the – say it aloud with us – "O!"-*chrolechia* species for their thick-rimmed apothecia, like Os all over the lichens.

SIMILAR SPECIES

O. tartarea is a similarly large species with conspicuous, thick-rimmed apothecia, but the thalli are consistently and thickly warted (tartareous – as in the calculus growth on unbrushed teeth!) and the discs are orange pink; it is found on acid bark of old trees, upland boulders and in mountain heaths.

THALLUS PALE – WHITISH OR GREY TO GREEN; APOTHECIA ROUND

BD 9

Crustose lichens

Lecanora chlarotera Rim Lichen

HOW TO SPOT Look for small whitish patches on smooth bark of twigs, branches and young trees.

DESCRIPTION **Crustose**. Thallus whitish, usually thin, up to 5cm, but generally smaller, sometimes stretched horizontally as tree girth expands. *Reproduction:* **Apothecia** abundant and often denser centrally, with pale brownish discs and whitish margins (). The margins are the same colour as the thallus and are minutely irregularly lumpy in outline () due to relatively large crystals within, which deform them.

WHERE On smooth bark of deciduous trees, throughout Britain and Ireland.

NOTES Several related species in the *L. chlarotera* complex are among the commonest Rim Lichens on bark. There are species of Rim Lichens in almost every sort of habitat (p. 82), and microscopy is usually needed to distinguish them. A few species have been selected on the facing page that might be identified fairly easily. Clues that you have different species include the colours of discs, and the thickness and evenness of margins. To be certain, however, microscopy is needed to examine the distribution of tiny crystals within the apothecia, along with chemical tests (spot tests).

SIMILAR SPECIES

Tephromela atra (p. 204) has similar fruits but with coal-black discs. The larger apothecia of *Pertusaria hymenea* (p. 221), with thicker irregular margins, can be confusing when they are moist and the discs are expanded and visible. ***Glaucomaria carpinea*** (**A**) often grows with *L. chlarotera* but can be distinguished by whitish discs due to a thin covering of fine crystals (pruina, 🔍). On stone and pavements, watch for the whitish fan-like marginal zone (🔍) of the pale grey thallus and dark brown discs of *L. campestris* (**B**), or blackish discs on a whitish thallus of *L. gangaleoides* (p. 204). On mortar, look for tiny discs (visible with 🔍 only) with no thallus of the ***Myriolecis dispersa*** group (**C** and **D**). This group is made up of several species, which all have jam-tart type apothecia with rough white margins, mostly without a visible thallus.

Lecanora polytropa Granite Speck

HOW TO SPOT Pale yellowish green stains on silica-rich rocks.

DESCRIPTION Crustose. Thallus pale yellowish green, flat and smooth, but sometimes obscured by the apothecia or lacking an obvious thallus. *Reproduction:* **Apothecia** are abundant, pale yellowish green to beige, with prominent margins the same colour or paler than the thallus when young (). Discs become convex and the margins become less visible with age. Apothecia have a smooth waxy texture, described as if carved out of jade ().

WHERE On silica-rich rocks, walls and sometimes sawn wood.

NOTES The thallus of this species can vary from barely there to thick and of separate thallus islands (areoles), so check for the young apothecia with neat, pale margins. The distinctive colour is due to the presence of usnic acid, the same chemical that makes *Usnea* (p. 68) yellowish green.

SIMILAR SPECIES

Often occurs with a suite of other yellowish-green *Lecanora* species on silica-rich rock, including *L. intricata* (**A**), with a visible thallus edge of delicately lobed areoles surrounding peripheral apothecia (). *L. sulphurea* has the same usnic colour and begins life as a parasite on other lichens (especially on ***Tephromela atra***, **B**) and has irregular, usually darkened apothecia when mature () with a well-developed yellowish thallus.

208 The pathways

Ophioparma ventosa Alpine Bloodspot

HOW TO SPOT Unmistakable large crust on boulders, with blood-red apothecia.

DESCRIPTION Crustose. Thallus grey to bluish grey or yellowish grey, sometimes pinkish, and then with a reddish growing edge. Up to 12cm across, thick, with a warted surface (). *Reproduction:* **Apothecia** deep red, numerous, rounded, with a paler margin the colour of the thallus, up to 2.5mm. These may become irregularly shaped and the margins can become less obvious on older apothecia.

WHERE On silica-rich boulders and scree in the uplands in sunny, exposed situations.

NOTES Two colour forms of this lichen occur, sometimes side by side. Yellowish thalli have the chemical usnic acid in the cortex, a sunscreening compound. It is often found with the tiny black dots of a fungal parasite which is specific to this lichen ().

SIMILAR SPECIES

Haematomma ochroleucum also has blood-red apothecia, but it is restricted to dry vertical or underhung rock faces including gravestones, and its entire surface is covered with powdery soredia.

Lecidella elaeochroma   Lecidella Lichen

HOW TO SPOT   Look out for this as part of mosaics on smooth bark along with *Lecanora* species (p. 206); it is often slightly greenish with black fruits.

DESCRIPTION   Crustose. Thallus grey to greenish, thin, smooth to slightly bumpy; greener in sunny positions and with a black or blue-black outline when in mosaics. ***Reproduction:* Apothecia** blackish to very dark bluish-black (), flat, with a raised margin when young, becoming convex later, up to 1mm.

WHERE   On smooth bark, especially on twigs and small branches.

NOTES   When bluish colours are present in the apothecia, it can be identified with confidence by eye at ; however, this species is vexingly variable, and not all specimens will be identifiable with confidence without a microscope. It is very common and should be suspected in most smooth-bark situations, particularly where there is a bit of nutrient input.

SIMILAR SPECIES

Fuscidea lightfootii (**A**) is also found on smooth bark, but it is made up of tiny convex islands (areoles), and has typically neatly rounded thalli, usually with mounds of powdery soredia (). On rocks, check the 'Lecidea' Group (p. 83). On old oaks in the south, look for the larger apothecia with a prominent margin and greenish dusting on their surfaces of ***Cresponea premnea*** (**B**). It has yellow pigments if you scratch the thallus.

Mycoblastus sanguinarius Bloody-heart Lichen

HOW TO SPOT Look for obvious black spots on a bluish-white thallus on old birches in the uplands.

DESCRIPTION Crustose. Thallus greyish white to bluish, up to 10cm, relatively thick and irregularly warted, and uneven ( ).
Reproduction: **Apothecia** conspicuously variable in size, up to 2mm, convex, black, with the margins often not obvious; when apothecia are broken off, a bright red spot ( ) is left on the thallus.

WHERE On acid bark, especially old birch and pines, or on silica-rich rocks or wood.

NOTES An abundant species in humid parts of upland Scotland and often found along with *Sphaerophorus globosus* (p. 118), *Parmelia saxatilis* (p. 163) and *Cladonia polydactyla* (p. 105).

SIMILAR SPECIES

No other lichen with convex black discs leaves red marks on the thallus where apothecia are scraped or worn away.

Rhizocarpon reductum Little Brown Map Lichen

HOW TO SPOT This is a pioneer of the lichen world, appearing as mouse-brown coloured spots on newly exposed stone and growing concentric rings of tiny black fruits.

DESCRIPTION Crustose. Thallus brownish-grey, 1–2cm wide, but sometimes growing together, cracked into flat, smooth tiny islands (areoles), with a black growing edge (). ***Reproduction: Apothecia*** black, about 0.5mm, with a black raised margin and black disc (), often arising in concentric rings on the thallus and separate from areoles.

WHERE On smooth silica-rich rocks, gravestones and monuments, but not at high elevation.

NOTES The commonest non-yellow Map Lichen. Look for the tidy apothecia and neat little thalli. Try to appreciate and examine (but maybe not yet identify!) other small lichens with black apothecia and black margins on rocks.

SIMILAR SPECIES

The similarly small ***Buellia aethalea*** (right, with *R. reductum* left) occurs in similar habitats and is similar colours, but the areoles and apothecia are even smaller, and the apothecia arise within areoles and push a tiny pale flap of thallus to the side as they develop (). Watch out for the Wine-gums *Lecidea* and others with black apothecia, common on rocks (see Baker's Dozen, p. 83).

Thelotrema lepadinum Bark Barnacles

HOW TO SPOT Look for dark craters on a creamy thallus in western woods; look closely and you'll see a raised flap of tissue inside the crater-like pore.

DESCRIPTION **Crustose**. Thallus white to creamy, greenish in shade, mostly smooth, irregularly expanding into large patches often over 5cm or more. *Reproduction:* **Apothecia sunken** below rings or flaps of thallus tissue, with an additional papery ring inside (🔍). Disc dark, but usually well hidden.

WHERE On smooth bark of deciduous trees and rarely on rocks. Abundant in western woods.

NOTES A particular suite of lichens can be found on smooth bark in temperate rainforests, and this is one of the commoner species, often along with Scripts and Commas (p. 84).

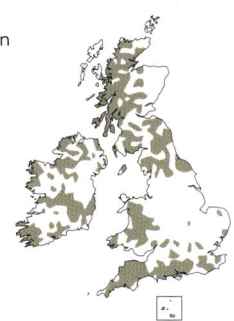

SIMILAR SPECIES

Watch out for *Lecanora* species that have been slug-grazed; molluscs often scrape the insides out of apothecia to get at the nutrient-rich spores. *Pertusaria hymenea* (p. 221) can look similar, but is usually greener and lacks an internal flap of tissue (🔍).

Crustose lichens

Icmadophila ericetorum Candy Lichen

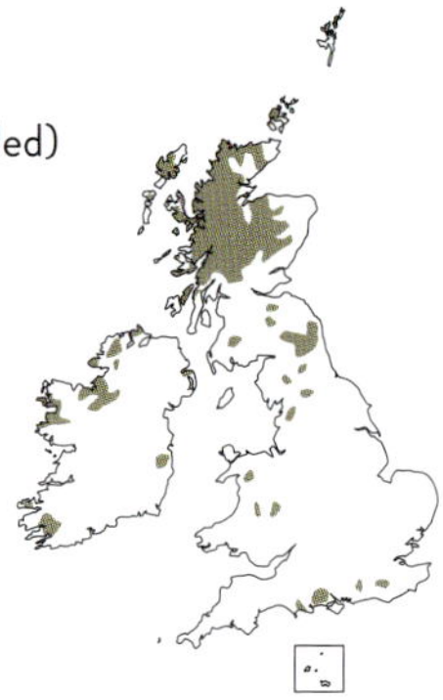

HOW TO SPOT Pretty pale bluish (dry) or minty green (wet or shaded) extensive thallus with flat pink discs (the eponymous candy) on very decayed wood or peaty soil, especially in the north.

DESCRIPTION Crustose. Thallus bluish green or minty green when damp, irregularly warted, extensive, up to 20cm or more. *Reproduction:* **Apothecia** pink, flat and very slightly raised from thallus ( ), up to 3mm, becoming irregular with age. Sometimes with a faint covering of crystalline powder (pruina), so appearing with a whitish bloom.

WHERE On rotten wood and peaty soil.

NOTES There is only one British species.

SIMILAR SPECIES

Look out for **Dibaeis baeomyces** (p. 222) on peaty soils, which also has pinkish apothecia, but these are convex and obviously stalked ( ). On thalli without stalked apothecia, be sure to look for tiny paler, whitish, smooth, rounded to flattened scale-like plates, which dislodge easily and are characteristic of *Dibaeis*.

Porpidia tuberculosa Cigarette Ash Lichen

HOW TO SPOT Large, pale to ashy grey smooth thalli on silica-rich rock, from garden walls and city monuments to mountain boulders.

DESCRIPTION Crustose. Thallus ashy grey to bluish, to about 10cm, thin and very flat, cracked into tiny islands (areoles,), finely patterned with darker and paler patches within; a pale outer zone is often visible from a distance. Mosaic forming, with dark grey to black outer growing edge. *Reproduction:* **Very fine powdery soredia** are produced in really tiny dot-like, dark grey patches that look like fine ash (soralia) up to 0.5mm wide, only visible with a 10× hand lens, flat in profile.

WHERE On silica-rich rocks, walls and monuments throughout, common.

NOTES This species seems to turn blue or purple on some rock types, so if you spot a flat thin thallus on rock with a bright colour, check for the distinctive, tiny round and flat soralia, thought to look like tuberculosis in lung tissue when first described.

SIMILAR SPECIES

On silica-rich rocks, look out for the whiter *Glaucomaria rupicola*, also a mosaic former, but with pale borders around thalli. It has inconspicuous pale apothecia covered in white crystal dusting (pruina) and white margins ( ).

Crustose lichens

Ochrolechia androgyna Powdery Saucer Lichen

HOW TO SPOT Large creamy white, very lumpy thalli often overgrowing mosses on old trees or on rocks. Look for creamy-yellowish spots of coarse soredia.

DESCRIPTION Crustose. Thallus whitish to pale creamy to 15cm, strongly irregularly thickened and warted, often with fan-like white growth at the edges, where it grows over mosses or other lichens. *Reproduction:* **Coarse, granular soredia** are yellow-tinged ( ) and in prominent convex clusters (soralia), later becoming widely spread across the central parts. Apothecia are occasional in Scotland or in southwest England.

WHERE On acid-barked and especially old mossy trees, or on silica-rich rocks and walls in nutrient-poor environments.

SIMILAR SPECIES

A number of lichens form large white crusts on trees, so always check their reproductive structures to tell them apart. Look out for **O. tartarea**, a strongly warted crust which produces abundant and large apothecia with orangish discs but without soredia; *Lepra amara* and *L. albescens* (opposite) have pale thalli with coarse soralia, lacking creamy or yellowish tones. On rock, see *Glaucomaria rupicola* (p. 215).

Lepra amara (Pertusaria amara) Bitter Wart

HOW TO SPOT Pale greyish to green patches on trees with white spots of coarse soredia.

DESCRIPTION Crustose. Thallus pale-grey to green-grey (greener in shade), usually thick, waxy, up to about 12cm, often with a white zone or concentric white lines near the growing edge (). **Reproduction: Coarse soredia** are formed in white dot-like to convex mounded patches (soralia) up to 1.5mm across (), which often contrast with the thallus.

WHERE On broadleaved trees and rarely on sheltered, humid rocks.

NOTES The common name reflects a useful characteristic of this lichen; if you haven't tasted it, moisten your fingertip to collect a few soredia and try it. The bitter taste isn't immediate, but it is diagnostic – and potent!

SIMILAR SPECIES

Lepra albescens is usually the same colour throughout and is not bitter tasting. *Pertusaria pertusa* and *P. hymenea* (p. 221) are pale greenish or greyish, but do not have powdery soredia (). *Ochrolechia androgyna* has yellowish mounded soredia (opposite).

Crustose lichens 217

THALLUS PALE - WHITISH OR GREY TO GREEN; SOREDIA

Lepraria incana Common Dust Lichen

HOW TO SPOT Irregular patches of dull bluish-green powdery soredia in shaded places, out of the way of direct rain, on trees, rocks or soil.

DESCRIPTION Crustose. Thallus bluish green, comprising compact powdery grains of soredia, creating fluffy masses with no other features or typical lichen layers, often spreading irregularly and widely. *Reproduction:* **Minute fluffy soredia** make up the entire thallus ( ).

WHERE On silica-rich rock, trees and soil out of the way of direct rainfall. This species is widespread in Britain and Ireland and beyond, tolerant of pollution and found even in cities on trees, walls and monuments.

NOTES Many Dust Lichens avoid direct wetting and are so hydrophobic that a droplet of water will remain perfectly round on their surfaces. *Lepraria* gives its name to the term 'leprose', referring to lichens made up of only soredia.

SIMILAR SPECIES

There are many Dust Lichens, so look for the distinctive bluish colour of this, along with its compact granules. In clean-air areas, there are many other species of soredia-only lichens, but most are not bluish. When you are starting out, just recognise them all as *Lepraria* species and move along!

218   The pathways

Lepra corallina (Pertusaria corallina) White Coral-crust

HOW TO SPOT A bright, pure white lichen forming large patches on silica-rich rocks.

DESCRIPTION Crustose. Thallus bright white or pale greyish white, with a zoned white edge forming patches up to 20cm, sometimes becoming very thick and lifting off the rock, occasionally breaking into thick blocks. *Reproduction:* **Peg-like, cylindrical isidia** uniformly cover central parts of the thallus (), later becoming branched repeatedly (only visible when the thallus is broken).

WHERE On silica-rich, dry rocks in exposed situations on walls, monuments, bridges etc., mainly in the uplands.

SIMILAR SPECIES

See the Supersized Crusts (p. 79) for other large white crusts. Several of those have coarse soredia and require chemical tests for identification, but only this one has white isidia. *Glaucomaria rupicola* (p. 215) is another white lichen often encountered on silica-rich, exposed rocks, but it is mosaic-forming, and rather smooth, with its apothecia discs at the same level as the thallus and hidden by a whitish crystalline covering (). Especially on coastal rocks, watch out for **Pertusaria pseudocorallina**, with a creamy beige thallus and with brown-tipped isidia ().

THALLUS PALE – WHITISH OR GREY TO GREEN; ISIDIA

BD 9

Ochrolechia frigida Arctic Saucer Lichen

HOW TO SPOT Whitish patches with spiky projections overgrowing plants on high moors and mountain summits.

DESCRIPTION Crustose. Thallus whitish to tinged yellowish, irregularly warted, covering soil and low-growing plants and mosses, later developing gently upward-arching spines like tiny fishbones (🔵). *Reproduction:* **Apothecia rare**, forming flat, smooth apricot-coloured discs with whitish margins up to 5mm (🔵). Occasionally found with yellowish, coarse soredia in convex clumps (soralia, 🔵).

WHERE On high moors and mountain summits, and at lower elevations further north, overgrowing mosses and low heath vegetation.

NOTES This is a real arctic-alpine species, found across the Arctic, Antarctic and subantarctic as well as high mountains in Europe. Although 'Fishbones and Apricot Tarts' is not an official common name, it is wonderfully evocative of the spine-like projections and the distinctively coloured apothecia.

SIMILAR SPECIES

O. androgyna (p. 216) also has coarse yellowish soralia, but lacks spiny projections.

Pertusaria pertusa Pepper Pot

HOW TO SPOT Pale grey or greenish extensive patches on bark with pronounced warts.

DESCRIPTION Crustose. Thallus pale grey to greenish grey, up to about 10cm or more, waxy, smooth or warted. ***Reproduction:* Apothecia** sunken within warts; warts are large, up to 2mm, constricted at the base and contain covered apothecia, each exposed only through its own small pores, usually 1–3 per wart (). These pinprick holes look a bit like the top of a traditional pepper pot.

WHERE On smooth to rough bark, sometimes on rocks.

NOTES These apothecia are a bit confusing, since the discs are almost completely enclosed. When they are moist, it is slightly easier to understand their anatomy, with several discs inside each wart more exposed as the warts expand and open.

SIMILAR SPECIES

P. hymenea (**A**) has a very similar thallus, but has a single apothecium per wart, and the pore is more open than in *P. pertusa* (**B**), particularly when wet, appearing like a thick-margined Rim Lichen. *Lepra amara* (p. 217) forms thick crusts, but has mounds of chalk-white soredia; its thallus can be greenish.

Baeomyces rufus Brown Beret lichen

HOW TO SPOT Look for large greenish-grey patches on rocks and soil, especially on rocky paths in damp shade.

DESCRIPTION Thallus greenish, thin, with a minutely lumpy texture or broken into rounded tiny islands (areoles, 🔍). **Reproduction: Apothecia** reddish to pinkish brown on minute and sometimes branching stalks, up to 6mm tall. When moist, the tips are paler and translucent.

WHERE On peat, silica-rich rocks, soil and gravel, in damp and shaded places, and common along rocky paths in the uplands.

NOTES Many completely unrelated fungi make stalked fruiting bodies. It is a good way to increase the chances of spores travelling to a new home.

SIMILAR SPECIES

Dibaeis baeomyces is similar and grows on soil, but prefers peaty situations, has a greyer thallus (when dry – greener when wet) and has rose-pink stalked apothecia, without brownish tones (**A**, 🔍). It is frequently found without apothecia but can be identified by paler, whitish, smooth, rounded to flattened plates very loosely attached and easily dislodged (**B**, 🔍), specialised for dispersal.

Arthonia radiata Asterisk Lichen

HOW TO SPOT One of the commonest species on twigs and smooth bark. Look for tiny pale white to creamy thalli marked with darker spots.

DESCRIPTION Crustose. Thallus comes in shades of white- to pale or fawn-grey or creamy. Up to 3cm across. Often forming mosaics, with pale thalli outlined by brown lines (👁). **Reproduction: Apothecia irregular to star-shaped**, up to 2mm, usually crowding the centre of the thallus, flat and without margins even when young (👁).

WHERE On twigs and smooth bark of branches and young trees.

NOTES This is a Comma Lichen, and the orange pigments of their particular algal photobiont (*Trentepohlia*, p. 84) may be visible (👁) with a light scratch of the thallus. Along with that so-called yellow to orange scratch, the apothecia are variable in shape but usually irregular in outline and lacking a raised or distinct margin – good clues that this is a very different group of fungi, unrelated to most lichens you will see.

SIMILAR SPECIES

There are many species of Comma Lichens with irregular apothecia without distinct margins. They are particularly diverse in western woods on smooth bark. Irregular dark brown to black apothecia and pale thalli on smooth bark is very likely to be *A. radiata*, but watch out for the more rounded – but still irregular – apothecia like tiny ink stains (👁) of *Diarthonis spadicea* (**A**), or the oval apothecia with red to pinkish crystalline dusting (pruina) of *Coniocarpon cinnabarinum* (**B**).

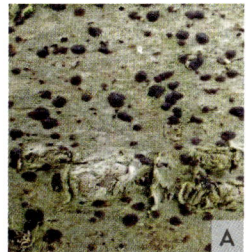

Graphis scripta Common Script Lichen

HOW TO SPOT Whitish thalli with short black lines, as if the fairies have been leaving tiny secret messages on the trees.

DESCRIPTION Crustose. Thallus whitish to pale greenish, smooth, with black lines. ***Reproduction: Apothecia form short, black lines (lirellae)***, sometimes branching (🔍); margins are lip-like and charcoal-black, sometimes sunken in the thallus and sometimes raised slightly above it; the disc within can be exposed or hidden when dry (when damp, the margins open more).

WHERE On smooth bark of trees on slightly shaded trunks and branches, especially in western woodlands.

NOTES This is a member of the Script Lichen family, so a gentle scratch of the thallus yields a yellow to orange mark (🔍) because of their orange-pigmented green algae (p. 84). There is wide variation in the apothecia in branching, openness and thickness of margins, corresponding to separate genetic lineages, but these can all be considered part of the *G. scripta* group.

SIMILAR SPECIES

G. elegans (**A**) has thicker apothecia margins, remaining closed, and with very small but consistent longitudinal grooves parallel with the main opening (🔍). Its thallus is slightly yellowish and lumpy (🔍) in comparison to *G. scripta*. Very crowded narrow

apothecia on a darker background (🔍) are likely to be ***Arthonia atra*** (**B**), but there are many species in the Scripts and Commas. Most require microscopy for identification. See also *Phaeographis dendritica* (opposite).

Phaeographis dendritica Starry Scribbles

HOW TO SPOT Striking whitish thalli with short grey lines forming a radiating pattern on smooth bark.

DESCRIPTION **Crustose**. Thallus pale grey to pale greenish, smooth, frequently powdery white, often with a radiating, star-like pattern of apothecia. ***Reproduction:*** **Apothecia form short, branching grey lines in a star-like pattern (lirellae)** level with the thallus (not raised); the disc is widely exposed and black, but covered thinly with white crystalline coating (pruina,). The ends of the lirellae are often sharply pointed (). The apothecia may make star-like patterns individually or collectively across the thallus.

WHERE On smooth bark of trees on slightly shaded trunks and branches, especially in the southwest where it can be locally common.

NOTES This is a member of the Script Lichen family, so a gentle scratch of the thallus yields a yellow to orange mark () because of their orange-pigmented algae (p. 84). This species is particularly conspicuous when it shows a radiating pattern of apothecia.

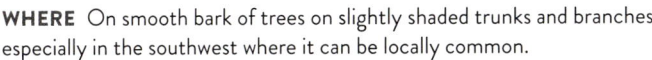

SIMILAR SPECIES

There are several *Phaeographis* and *Graphis* species in Britain and Ireland, and although microscopy is needed to identify most of them, the ones here can usually be recognised with just your hand lens.

Crustose lichens 225

Lecanactis abietina Old-wood Lichen

HOW TO SPOT Extensive patches of a distinctive pinkish-grey thin crust on the dry sides of old trees, visible from metres away.

DESCRIPTION Crustose. Thallus thin, pale grey typically with distinctive pinkish or mauve tones, covering large areas or present only in crevices of bark. *Reproduction:* **Characteristic regularly spaced white dots on the thallus** are protruding asexual spore-masses atop tiny black stalks (). At 10×, the white spore-masses are just visible on the tops of black pycnidia, but the pycnidia themselves are usually covered in thallus and only black when worn way. Apothecia sometimes also present, up to 2mm with a whitish or yellowish-grey fine crystalline covering (pruina) on the dark-brown to blackish discs ().

WHERE On the dry sides and rough bark of old trees, especially on old oaks and other deciduous trees, but also on conifers.

NOTES This species is restricted to relatively old woods, but is not exclusive to very old trees. Look out for the apothecia that can be quite camouflaged due to their heavy covering of fine crystals (pruina); they sometimes appear a bit yellowish (). The dry sides of rough-barked trees are home to a specialist community of lichens including *Chrysothrix candelaris* (p. 197) along with many other crusts.

apothecia

Pyrenula macrospora Flecked Pox

HOW TO SPOT Look for evenly spaced prominent black dots on smooth bark in the west.

DESCRIPTION Crustose. Thallus yellowish, greenish to orangish green, waxy, in mosaics on smooth bark, outlined in black and dotted with conspicuous black fruiting bodies and minute white flecks (pseudocyphellae,). *Reproduction:* **Perithecia** are black, closed spore-bearing structures up to 1mm wide, conical in shape, opening with a tiny pore ( ).

WHERE On shaded smooth bark of deciduous trees in the south and west.

NOTES Like the Scripts and Commas, the Pox Lichens are very diverse in the tropics, but also at home in the mild weather of the temperate rainforest. You may hear about all these smooth-bark lichens referred to as the smoothies. Often found with *Graphis scripta* (p. 224) and *Thelotrema lepadinum* (p. 213).

SIMILAR SPECIES

Other Pox Lichens are also found in Britain and Ireland on smooth bark in temperate rainforests, but none have perithecia as large as these.

Lichenomphalia ericetorum Heath Navel

HOW TO SPOT Look for a thin, even coating of rich forest green on rotten wood, peat or on moss – and then look for tiny yellow-brown mushrooms with delicately descending gills.

DESCRIPTION *Crustose*. Thallus deep emerald green, extensive, of evenly spread, tiny spherical globes ( ) – the symbiotic thallus of this lichen. At 10×, the spheres almost glow as light shines through. *Reproduction:* **Mushrooms** are up to 3cm tall, pale brownish-yellow to creamy white, with few, thick gills that taper down the stalks; the mushroom cap has a depression in the centre like a navel.

WHERE On well-rotted wood or peaty soil or moss in nutrient-poor situations.

NOTES A small number of lichens are formed by the mushroom-forming fungi, including this group of tiny mushrooms. Once you've examined the thallus of this with your hand lens, you'll soon start to see its deep green thallus on rotting wood or peat from a distance. On moss, it is probably easier to spot the tiny mushrooms.

SIMILAR SPECIES

L. alpina also has a thallus of tiny green globes, but occurs at very high altitude or latitude and has bright orange-yellow mushrooms (p. 18).
L. hudsoniana has yellowish-brown mushrooms, but a different thallus, of small rounded scales (squamules) with prominent margins (, arrows) growing on peaty soil or decaying vegetation.

Acarospora fuscata Brown Cobblestone Lichen

HOW TO SPOT Often found as extensive brown patches up to 20cm or more on flat rocks.

DESCRIPTION Crustose. Thallus rich chestnut brown, forming a thick crust, often visible at a distance when extensive; made up of tiny islands (areoles), each irregular and separate from the others, though often pushed together tightly, and raised at the edges like dried mud. *Reproduction:* **Apothecia**. Mature areoles have one to a few small, darker brown sunken areas, which are the apothecia.

WHERE On silica-rich rocks usually with some nutrient input, including farm walls, memorials, and roofing tiles or outcrops in agricultural landscapes.

NOTES An abundant species throughout Britain and Ireland. Often called the cracked mud lichen.

SIMILAR SPECIES

No other lichen forms extensive chestnut-brown patches cracked into areoles and with darker brown sunken apothecia. Other *Acarospora* species also have thalli with sunken darker apothecia in areoles, but are much less frequent and mostly much smaller (see *A. sinopica*, p. 233).

Verrucaria nigrescens Blackwart Lichen

HOW TO SPOT Small, matt black patches on limestone or mortar or cement.

DESCRIPTION Crustose. Thallus dark brown to dark olive to black, cracked into thick island-like areas (areoles, 🔍), sometimes with brownish isidia-like outgrowths at their edges (🔍).
Reproduction: Perithecia are mostly sunken within the thallus, resulting in darker centres of areoles with a tiny raised pimple and a pore at their tops for spore discharge (🔍).

WHERE On lime-rich mortar, paving slabs, limestone and monuments.

NOTES There are many species of *Verrucaria*, and they are difficult to ID – as a beginner, or even an intermediate lichen enthusiast, they are probably best left at the genus level, as microscopy and dedication are both required. This species can be learnt by beginners, but keep an eye out for other crusts with pimple-like perithecia like these, some with sunken perithecia, and some protruding.

SIMILAR SPECIES

In similar habitats, look out for the black crust *Placynthium nigrum* (opposite), which has a blue-black thallus margin (🔍) and is covered with isidia (🔍). *Hydropunctaria maura* (p. 232) is also matt black but occurs only on coastal rocks or where sea-spray is an influence.

THALLUS BROWN TO BLACK

Placynthium nigrum Blackthread Lichen

HOW TO SPOT Thick matt black crusts with subtle deep blue-black edge on lime-rich rocks.

DESCRIPTION Crustose. Thallus brown-black to jet-black, matt, to 12cm, of small, flat scales (squamules) with tiny lobes on their edges; has a distinctive blue-black growing edge. ***Reproduction:*** **Dense isidia** often cover central parts of the thallus, variable in shape; black **apothecia** are occasional with black, shining margins, concave to flat at first, becoming convex, usually less than 1mm.

WHERE On lime-rich natural and human-made surfaces, like limestone, concrete, compacted soil, stonework, mortar etc., especially in places slow to dry.

SIMILAR SPECIES

From a distance may be confused with *Verrucaria nigrescens* (opposite), which is also very dark and occurs on limestones and other lime-rich substrates such as mortar and concrete. It has a smoother texture ( ) and raised pimples of the tips of the otherwise hidden perithecia.

THALLUS BROWN TO BLACK

Hydropunctaria maura (*Verrucaria maura*) Tar Lichen

HOW TO SPOT Think it's just rock you're looking at? Look again; those extensive black patches just around the upper limits of seaweeds on coastal rocks are actually lichens.

DESCRIPTION Crustose. Thallus matt black, forming small patches or spreading widely and covering square metres, with very fine separate cracks throughout (👁). *Reproduction:* Perithecia are sunken in the thallus and visible only as tiny black pimples of even size (👁). Try to spot the very tiny pores at the tips of each one.

WHERE On seaside rocks, growing from below to well above the high-tide mark. Very common. Often extensive at the middle shore, just above seaweeds, and forming separate, round thalli in higher positions relative to the sea, often with the orange Firedots *Flavoplaca marina* and *Variospora thallincola* (p. 190) and Sunburst lichens (p. 139).

NOTES *Hydropunctaria* is a member of the 'Verrucaria' group, mostly rock-dwelling lichens with perithecia. When you find this species, be sure to look out for the minutely fruticose *Lichina* species (p. 96), which can occur in similar places.

SIMILAR SPECIES

Several black *Hydropunctaria* species are encountered on marine rocks, but *H. maura* is by far the commonest. Look out for waxy green round patches of **Wahlenbergiella mucosa** (shown here), another relative in the 'Verrucaria' group on lower shore rocks, only exposed at low tides and often growing with barnacles and limpets.

232 The pathways

Rusty Reds

Iron-rich rocks often have crustose lichens that are rusty red or orange. Check the growing edges of the thallus and the reproductive structures closely. You may not be able to be completely confident about these just by examining them with a hand lens, but you can propose some informed hypotheses about their identification.

Haugania (Rhizocarpon) oederi (**A**) produces small, deep reddish thalli with small black apothecia, which are convex and irregular on their upper surfaces, making the apothecia look messy under the hand lens (🔍).

Tremolecia atrata (**B**) has small, deep reddish-orange thalli with small black apothecia. When they are young, the apothecia are smoothly concave, and they look neat under a hand lens (🔍).

Acarospora sinopica (**C**) produces deep reddish-orange thalli with clearly separate areoles, each with 1–3 darker orange, sunken, but inconspicuous apothecia (🔍).

Several *Lecidea* (**D**) or *Porpidia* species sometimes develop rusty colours on iron-rich rock. Both genera typically have black apothecia with black margins. *Lecidea* species often have flat apothecia, while some *Porpidia* have convex apothecia (🔍), but microscopy is needed to identify most of them confidently, even to genus.

Porpidia melinodes (**E**) has tangerine-coloured thalli, often with a paler grey border and tiny, neat, greyish clusters of soredia (🔍).

Porpidia flavocruenta (**F**) has tangerine-coloured thalli with black apothecia with black margins above the thallus surface. The apothecia sometimes have a grey dusting of crystals on the disc (🔍).

THALLUS REDDISH TO ORANGE

Crustose lichens

TROUBLESHOOTING

First, a gentle reminder that not every lichen you find will be able to be identified, by you or indeed perhaps by anyone. Sometimes, a particular lichen is immature or in a particular place has a very unusual look, completely atypical for that species, and these can be difficult. If you get to the genus or group in the sense of the Baker's Dozen, you may want to consider the job done.

Nevertheless, a few potential common stumbling blocks are dealt with below.

What is *not* a lichen?

Sometimes when you are getting started it can be difficult to know what *is* a lichen and what is not a lichen!

Take a closer look and ask yourself:

- **Is it biological?** That is, does it have repeating patterns of growth or development? No? Probably paint. We've all been fooled once or twice; keep looking!

Yellow paint on a limestone wall, irregular in shape and very smooth when magnified.

- **Is it a bryophyte – i.e. a moss or a liverwort?** Mosses and liverworts are little plants, often with tiny stems and leaves, and often green. Not all mosses and liverworts have regular patterns of leaves and stems, nor are they all green, but you'll get the hang of it.
- **Is it fuzzy when you look really closely?** Orange-fuzzy or green-fuzzy is probably just free-living algae. These look fuzzy because the colonies are made of microscopic parts grouped together, and difficult to see by eye or even with a 10× lens. In lichens (composed of both fungi and algae or cyanobacteria) the organisation of the upper layer tends not to produce a fine thread-like structure (though there are exceptions you are unlikely to encounter in Britain and Ireland!).

Some liverworts are green and simple, without leaves (**A**, *Metzgeria* sp.), while others have minute leaves and stems (**B**, *Frullania* sp.).

Trentepohlia (left and middle) is a filamentous free-living green alga but with strong orange pigments and can be found in a huge variety of settings from mountain rocks, woodland trees, and even old walls, but it frequently produces diffuse and extensive orange colonies, often visible from a distance. In nutrient-enriched and polluted settings, free living green algae (right) can be abundant. Their fuzzy texture of fine filaments can be seen with a 10× hand lens.

Colour

Colour is of course subjective and the same colour will be perceived and described differently among observers. Although there are expected differences among lichens, there is a wide range of colour possibilities even within the same species. A variety of factors can interact, including degree of light and shade, age of specimen, infection with other fungi, colonisation by epiphytic algae (free-living algae growing on top of lichens), genetic variation, and degree of wetness, to name a few.

Green

The amount of sun or shade a lichen receives while growing triggers a physiological response regulating the amount of UV light that can penetrate its outer layer. In effect, lichens make sunscreening chemicals in the upper cortex (or exposed parts of soredia), and high UV levels trigger more production of these, protecting the photobionts underneath. The shadier the conditions, the less sunscreening is needed, and therefore the lichen will appear greener. *Xanthoria parietina* is bright yellow orange in exposed situations on sunny twigs, but greenish blue on the shaded side of branches (see p. 139). Likewise, shade-growing lichens like the Shield Lichens and other green-algal species are more greenish, as fewer pigments are produced to shield the photobiont layer beneath the cortex.

Yellowish green is a very significant colour for lichenologists, and it is because of a single compound called usnic acid – the chemical that gives *Usnea* its distinct yellowish green colour. This is also found in *Flavoparmelia* species, some *Xanthoparmelia* and a handful of *Cladonia* species, among

The *Flavoparmelia* on the right has usnic acid, making it yellowish and helping it stand out against the greyer colours of *Parmelia* (left) and *Hypotrachyna* (top middle).

Crustose lichens can have usnic acid, too. The *Lecanora polytropa* in the centre contains the subtler usnic acid yellow, and contrasts with the brighter yellow of *Candelariella vitellina*.

others. It is best seen and understood in daylight, not under indoor lighting, which can obscure its distinctive hue. When you are starting out, it isn't always easy to distinguish this colour, so we have included lichens containing usnic acid within the larger group of pale whitish or grey to green colour palette. Sometimes the same lichen species can be different colours because of the presence or absence of usnic acid, as seen in Alpine Bloodspot *Ophioparma ventosa* (p. 209).

Brown and black

Melanin is a group of chemicals common in fungi and animals – you will recognise the name as being associated with changes in human skin due to exposure to the sun. Fungi and animals both can produce more protective melanin in high-UV situations. In lichens, UV-induced darkening due to melanin happens across the exposed thallus surface. *Platismatia glauca* can be dark brown on fenceposts in full sun, and greenish grey in shaded woods.

The age and condition of lichens will also cause differences in appearance, with older parts often showing discolorations due to local tissue death, infection or colonisation. Black and brown spots are often local infections or dead patches of tissue. Many different fungi live on and in lichens, including those that are specialist parasites of lichens.

Pink and red

There are various explanations and causes of rusty orange to red or pink colours. Pink and red spots are often signs of local tissue death or infection.

There are also a handful of common fungi that specialise in living on lichens (lichenicolous fungi) which are pinkish colours, like *Illosporiopsis christiansenii* (a few distinct, shocking-pink blobs up to about 1mm on *Parmelia*, *Physcia* and others), *Marchandiomyces corallinus* (many small pink blobs on bleached lichen parts, especially of *Parmelia*), the orangish-pink *Erythricium aurantiacum* (many small orangish-pink blobs on *Physcia*), or *Laetisaria lichenicola*, which turns *Physcia* pink without blob-like structures.

Crustose lichens can change the chemistry of the rock they grow on, with some allowing the iron in the rocks to come to the surface and rust on contact with air and moisture (see p. 11).

Blue and purple – mountain crusts

We don't know what causes this, but you may spot a blue or purple crustose lichen in the mountains. Check *Porpidia tuberculosa*, which is a common species that has these colours, possibly on particular rock types. *Fuscidea lygaea*, although not described in this book, is also fairly common in upland areas and often has a purplish colour a bit like weak blackcurrant juice!

Baby lichens

Everybody has to start somewhere, and lichens do too. If a lichen isn't yet reproducing, it is probably not mature, and therefore its typical structures needed for ID may not be visible. The best advice is to look around for a better, more mature, specimen. Failing that, let it grow, and come back in a year or two to look again later. Juvenile lichens can usually only be identified by people already familiar with that species after quite a bit of experience.

Pseudocyphellae

Pseudocyphellae are actually pores (for airflow) in the surface of a lichen (upper or lower surface, see images pp. 38 and 165). They are white for two reasons: 1) there are no algae there, and 2) the cortex thins and eventually opens, revealing the (usually) white medulla within. *Parmelia* illustrates the classic example of pseudocyphellae, forming a distinctive fine network pattern on lobe tips, which open to the medulla as they develop.

Bright red spots on *Parmelia* species are often places where the tissue has died, and the breakdown of the chemicals within (salazinic acid) causes a bright red spot to appear.

Several lichen parasitic fungi produce pink to red colours including the orangish-pink *Erythricium aurantiacum* on *Physcia*.

However, some lichens have white patterning that doesn't allow airflow, which can be confusing. For example, *Platismatia glauca* (p. 172) has tiny white areas on the upper surface because the algae are not distributed evenly, and the cortex is very thin. Look with your hand lens, and you'll see that the shiny cortex is continuous across the top of all the white patches. *Parmotrema reticulatum* (p. 171) is a rare species that has a similar surface patterning which is slightly more distinctive (but a less common lichen!).

Reproductive structures

Can't find any reproductive structures on your lichen?
- Look again at all the different parts of the thallus to see if you can locate some – it may be that there are only one or two!
- If you still can't find any, look around for another example of that lichen which does have some more obvious reproductive structures. It may be that the first one you looked at had had its fruiting bodies eaten by a slug, or that the lichen is very young and hasn't produced any reproductive parts yet.

Soredia or isidia?
- Sometimes eroded or knobbly isidia can look like coarse, granular soredia. Look again and try to find more examples of the structure. Look carefully at a lot of branches/lobes, both young and old. Can you see how reproductive parts form and develop on young branches? In some species with coarse soredia, it can be difficult to tell which category you might describe them as, but remember that soredia arise from within a thallus, whereas isidia typically arise from the cortex itself. Find the youngest examples and try to understand their development.
- In a few cases like *Melanelixia subaurifera* and many *Usnea* species, isidia-like structures arise within soralia – openings in the cortex, meaning that the cortex ends at the base of each.
- What is their distribution and density, their shape in profile and outline? Keep using your hand lens to check the reproductive structures of lichens, and you'll start to see differences in the distribution of soredia and isidia. These usually correspond to different species. The more you practise looking, the more you'll begin to understand these differences.

Still stuck?
Found something in a place not shown on the map? The maps cannot show individual records, but you can always check detailed maps at the British Lichen Society website: Species and maps. For anything else, check the sections on **Resources and next steps**, and **Further reading** (pp. 239–43).

RESOURCES AND NEXT STEPS

Have patience, be kind to yourself and keep looking carefully. Using a hand lens is crucial to progressing. Once you've begun your lichen journey, consider joining an online or local group; find out more on the Local Groups and Online Learning sections of the British Lichen Society website. It makes a huge difference to learn with others.

Hand lenses

A small, folding hand lens with ten times magnification is all you need to get started; more magnification often comes with more distortion unless you pay more. It's better to spend money on a better lens than two (or more) inferior lenses. Consider buying from a specialist company like NHBS, who will have already done the research about the quality of a product. Kite or Opticron both make quality lenses.

A good lens with about a diameter of 20mm should not be restrictively expensive, but doublet or even triplet lenses are well worth the investment, and some even have good LED lights included.

A good folding lens. There are many inexpensive and poor-quality lenses for sale online but you will see more, and see more clearly, with a well-chosen lens.

RECOMMENDATIONS FOR YOUR KIT BAG

- This field guide!
- **Hand lens:** we recommend magnification of 10×, and a lens with a built-in LED light will make a world of difference
- **Camera/phone.** Try holding your hand lens in front of your phone camera lens for close-up images.
- **Notebook and pencil** (more waterproof in wet weather)
- **Something waterproof to sit on,** and/or **knee pads**
- **Weather-appropriate gear** – it can get chilly, really chilly when your pace of movement slows down to lichen speed, so think about layers and even fingerless gloves in cool weather
- **Snacks/flask** (brain food!)

As your lichen hunting habit develops, your kit list will mature as well, but this is a good starting list. See the section on Collecting lichens (p. 241).

Online information and photographs

Being able to check your identifications is useful, and there are many excellent websites with collections of images organised in various ways. Be sure to spend some time browsing the British Lichen Society website, too.

British Lichen Society (BLS) britishlichensociety.org.uk – an invaluable resource for anyone interested in lichens. Includes learning resources, identification tips and species descriptions as well as lists of events, ways to get involved, lots of links and information about how to submit records. Authoritative and complete family-level treatments with keys and descriptions and a glossary are provided at britishlichensociety.org.uk/identification/lgbi3.

British Lichens britishlichens.co.uk – photographs of many common species with up-to-date taxonomy and a really useful picture index using sets of thumbnails, in addition to a handful of clever habitat tours with photos; maintained by the BLS.

Cumbria Lichen and Bryophyte Group cumbrialichensbryophytes.org.uk – trip reports and blogs for the Cumbria local group, including an especially useful set of mapping resources for Cumbrian species.

Dorset Lichens dorsetnature.co.uk/Dorset-lichen.html – a richly illustrated website including thumbnails for quick identification and up-to-date taxonomy, organised by growth forms. Accounts include species descriptions and images, including of microscopic structures. Handy lists of lichens from some local sites.

Fungi of Great Britain and Ireland fungi.myspecies.info – a complete list of fungi including lichens with up-to-date taxonomy and descriptions including photographs of specimens in the field, or sometimes from preserved collections, and often with excellent-quality images of microscopic features.

Irish Lichens irishlichens.ie – a richly illustrated website including thumbnails for quick identification and up-to-date taxonomy, organised by growth forms and with useful illustrated guide to growth forms and fruiting bodies on landing page. Accounts include species descriptions and images including microscopic structures.

Italic 8.0 – The information system for Italian lichens italic.units.it – with pictorial keys leading to descriptions and distribution maps for Italy. An expansive and useful resource, and see especially the interactive guide to lichens, which includes lots of images and a substantial overlap with British and Irish lichens.

LichenIreland habitas.org.uk/lichenireland – distribution maps and descriptions with photos of some species.

Lichens Marins lichensmaritimes.org – dedicated to the lichens of Brittany, with a large degree of overlap with British and Irish species, with descriptions and photographs.

The OPAL Lichen field guide dryades.units.it/home/pdf/OPALlichenfieldguide.pdf – a user-friendly key to over 70 lichens on trees, targeted to England, but works well in many parts of Britain and Ireland; also available through **dryades.units.it** online.

Scottish Lichens scottishlichens.co.uk – a habitat-based introduction to lichens in Scotland, great for beginner-level learners.

Stridvall Lichens gallery stridvall.se/lichens/gallery – a stunning collection of images from Sweden with substantial overlap with British and Irish species.

University of Oslo Lichen Herbarium: macrolichens project nhm2.uio.no/botanisk/lav/LAVFLORA/LAVFLORA.HTM – with keys, descriptions and images of macrolichens in Norway, in Norwegian, with substantial overlap with British and Irish species.

University of Oslo Lichen Herbarium: species gallery nhm2.uio.no/lav/web/index.html – a rich image library with substantial overlap with British and Irish species.

Wales Lichens wales-lichens.org.uk – focusing on the conservation of lichens in Wales.

Ways of Enlichenment waysoflichenment.net/lichens – North American lichens by morphological grouping, by genus or by higher-level taxonomy, including thousands of wonderful images.

Platforms for help with identification

British Lichen Society britishlichensociety.org.uk runs online meetings for absolute beginners – the LABS groups. See the Zoom Groups and Online learning section of the Learning pages on the website.

iNaturalistUK uk.inaturalist.org is an online network of people sharing biodiversity information and is a great way to start recording lichens, as the algorithms that platform uses allow users to get suggestions for identification. The automated suggestions are not always right, but they may help you get close, and the online community can help check your work.

Lichens Connecting People – a worldwide Facebook group for lichens.

UK FungiForum fungi.org.uk – for help with identification of fungi, including a lichen forum.

Courses

Courses are offered by the Royal Botanic Garden Edinburgh, the Field Studies Council and the British Lichen Society, so check their websites for more information.

Recording lichens

Over the past few decades, the British Lichen Society (BLS) has gathered a huge amount of information about the distribution of lichen species. Its database now contains more than 2 million records, all with details of location, date, recorder and species, and for species of interest there is often further information on substrate, position and identification. The LichenIreland project has also done much to add to our knowledge of lichens in Ireland. These records are used to generate range maps, understand threats and conservation needs, and for scientific research of all sorts.

We invite you to record the lichens that you see in this guide. Visit the BLS website Recording pages to find out more. We have mostly chosen species that are straightforward to identify, but if you only feel confident to record one or two species at first, that's fine; reflecting an oddity of human behaviour, there is often an overemphasis in many recording datasets of rare species and not enough records of common ones, so please consider starting to send in your records. To increase your confidence, go on a course or join a local group to consolidate your species concepts!

Collecting lichens

The species in this book don't need to be collected for identification. If you later decide you want to begin to assemble your own reference collection, see the BLS website for more information on collecting. If you do decide you need a small piece for checking at home, be sure you collect into paper packets with the following minimum information: date, location, habitat and substrate, and please don't collect the only example of something you've spotted – it may be rare.

FURTHER READING

Books and other publications

Habitat-based fold-out guides

ACTON, A., GRIFFITH, A. 2008. Lichens of Scotland's Rainforest: Guide 1 – Lichens on ash, hazel, willow, rowan and old oak. Plantlife. – *a fold-out guide to species on nutrient-rich bark; a pdf can be found online at Plantlife.org.uk or Alliance for Scotland's Rainforest savingscotlandsrainforest.org.uk/resources-identification.*

ACTON, A., GRIFFITH, A. 2008. Lichens of Scotland's Rainforest: Guide 2 – Lichens on birch, alder and oak. Plantlife. – *a fold-out guide to species on acid bark; a pdf can be found online at Plantlife.org.uk or the Alliance for Scotland's Rainforest savingscotlandsrainforest.org.uk/resources-identification.*

DOBSON, F. 2004. Guide to Common Churchyard Lichens. Field Studies Council, Shrewsbury. – *a fold-out guide with photographs and a multi-access key to over 50 common species.*

DOBSON, F. 2006. Urban Lichens on Trees and Wood: Pt. 1 (WildID). Field Studies Council, Shrewsbury. – *a fold-out guide with photographs and a multi-access key to nearly 50 common species.*

DOBSON, F. 2006. Urban Lichens (2) on Stone and Soil: Pt. 2 (WildID). Field Studies Council, Shrewsbury. – *a fold-out guide with multi-access key to nearly 50 common species.*

DOBSON, F. 2007. Guide to Lichens of Heaths and Moors (WildID). Field Studies Council, Shrewsbury. – *a fold-out guide with photographs and a multi-access key to 60 common species.*

DOBSON, F. 2009. Rocky Shore Lichens (WildID). Field Studies Council, Shrewsbury. – *a fold-out guide with photographs and a multi-access key to over 60 common species.*

Plantlife. 2022. Lichens of temperate rainforest in the Lake District. Guide 1 – the *Lobarion* lichens of ash, hazel, willow, rowan and old oak. – *a fold-out guide to species on nutrient-rich bark; a pdf can be found online at Plantlife.org.uk.*

Plantlife. 2022. Lichens of temperate rainforest in the Lake District. Guide 2 – the *Parmelion* lichens of birch, alder and oak. – *a fold-out guide to species on acid bark; a pdf can be found online at Plantlife.org.uk.*

Plantlife. 2022. Lichens of temperate rainforest in Southwest England. Guide 1 – the *Lobarion* lichens of ash, hazel, willow, rowan and old oak. – *a fold-out guide to species on nutrient-rich bark; a pdf can be found online at Plantlife.org.uk.*

Plantlife. 2022. Lichens of temperate rainforest in Southwest England. Guide 2 – the *Parmelion* lichens of birch, alder and oak. – *a fold-out guide to species on acid bark; a pdf can be found online at Plantlife.org.uk.*

Plantlife. 2022. Lichens of temperate rainforest in Wales. Guide 1 – the *Lobarion* lichens of ash, hazel, willow, rowan and old oak. – *a fold-out guide to species on nutrient-rich bark; a pdf can be found online at Plantlife.org.uk.*

Plantlife. 2022. Lichens of temperate rainforest in Wales. Guide 2 – the *Parmelion* lichens of birch, alder and oak. – *a fold-out guide to species on acid bark; a pdf can be found online at Plantlife.org.uk.*

Plantlife. 2022. Lichens of temperate rainforest in Wales. Guide 3 – the *Graphidion* lichens of smooth-barked trees. – *a fold-out guide to species on smooth bark; a pdf can be found online at Plantlife.org.uk.*

WOLSELEY P., JAMES, P., ALEXANDER, D. 2003. Lichens on Twigs (WildID). Field Studies Council, Shrewsbury. – *a fold-out guide to 60 common species.*

Books

COPPINS, S., COPPINS, B.J. 2012. *Atlantic hazel: Scotland's Special Woodlands.* Atlantic Hazel Action Group. – *a detailed yet accessible book showcasing the importance of hazel woods in Scotland, including history, management and diversity of special lichens.*

DOBSON, F. 2013. *A Field Key to Lichens on Trees.* Richmond Press. – *a dichotomous key using field characters to lichens commonly found on trees; now available through the British Lichen Society as a pdf online britishlichensociety.org.uk/the-society/bls-shop with update notes including name changes and notes on species.*

DOBSON, F. 2013. *A Field Key to Common Churchyard Lichens.* Richmond Press. – *a dichotomous key using field characters to lichens commonly found on stone, worked wood, mortar, moss and soil in the lowlands; now available through the British Lichen Society as a pdf online britishlichensociety.org.uk/the-society/bls-shop with update notes including name changes and notes on species.*

DOBSON, F. 2013. *A Field Key to Coastal and Seashore Lichens.* Richmond Press. – *a dichotomous key using field characters to lichens commonly found on coasts and shores; now available through the British Lichen Society as a pdf online britishlichensociety.org.uk/the-society/bls-shop with update notes including name changes and notes on species.*

DOBSON, F. 2018. *Lichens: an Illustrated Guide to the British and Irish Species (7th edition),* Richmond Publishing and British Lichen Society. – *an indispensable intermediate-level field guide that introduces next steps, including keys and chemical tests; over 1000 lichens are included.*

LÜCKING, R, & SPRIBILLE, T. 2024. *The Lives of Lichens: A Natural History,* Princeton University Press. – *a richly illustrated exploration of lichen biology with stunning photography and engaging writing.*

WHELAN, P. 2024. *Lichens of Ireland & Great Britain: A Visual Guide to Their Identification,* Holm Oak Press. – *including descriptions and annotated photographs of over 700 species, with notes on chemistry and microscopy.*

GLOSSARY

Acid bark — bark with low pH and low levels of nutrients. Especially conifers, but also birch, alder and some oak trees – individual trees will vary in their pH.

Apothecia (singular: **apothecium**) — saucer- or cup-shaped sexual fruiting bodies of many ascomycete fungi, with open **discs** and often with raised **margins**.

Areole — an island-like lichenised structure, mostly in crustose lichens, either arising independently from a fungal-only early structure (prothallus) or by breaking into smaller parts as the crust develops, like old cracked paint or dried mud.

Ascomycete — a member of the large group of fungi that reproduce via sexual spores inside an ascus or sac.

Bullate — with raised, convex areas in the upper surface (e.g. *Peltigera membranacea*, p. 146).

Calcareous — high in calcium carbonate; calcareous rocks or soil are base-rich or lime-rich (alkaline), with high available cations such as calcium and magnesium; especially limestone, marble, mortar or cement.

Cephalodia (singular: **cephalodium**) — lichenised structures that contain nitrogen-fixing cyanobacteria, often different in colour and structure to the rest of the lichen.

Cilia (singular: **cilium**) — hair-like filaments that extend from the edge of lobe (10×) (e.g. *Parmotrema* and *Physcia tenella*, p. 171 and p. 180); contrast with **rhizines**, which arise from the lower surfaces of lobes.

Colour — one of the trickiest parts of lichenology to master! Colours are noted in the dry state and often change when wet. See basics of ID and troubleshooting, pp. 24 and 234.

Cortex — the outer layer of a lichen, made of thick-walled hyphae growing close together and bound with a gel-like polysaccharide matrix; it is a protective and absorptive structure, with tightly packed fungal threads with thick cell walls and chemical compounds which protect the algal layer from harmful UV light.

Cracked — a crustose lichen that develops as a continuous, expanding flat layer, and later develops cracks, sometimes breaking completely into separate areoles (e.g. *Ochrolechia parella*, p. 205).

Crustose (crusts) — a flat, crust-like growth form of lichens, in which the lower surface is completely attached to the surface it's growing on and cannot be removed without the substrate attached.

Cyanobacteria — photosynthetic bacteria that are bluish-green or dark emerald green in colour *en masse*; these are often able to fix carbon as green plants do, but also to fix atmospheric nitrogen into usable forms.

Cyanolichen — lichen with cyanobacteria as photobionts; often dark grey, brown or black in colour and usually found in microhabitats that are frequently wet, like mossy branches and tree bases, or limestone rocks.

Cyphellae (singular: **Cyphella**) — a neat pore in the surface of a lichen, often with a membrane, see *Sticta*, pp. 149–50.

Deciduous — a plant that loses its leaves in the winter.

Disc — the fertile layer in an apothecium including the spore-bearing cells (**asci**) and sterile tissues that support them; often coloured differently to the thallus of a lichen and sometimes contrasting with the margin.

Elongate (or **elongated**) — longer than wide, used to describe apothecia that can look like scribbles of writing (see description of lirellae) or lobe shape.

Epiphyte (or **epiphytic**) — a plant or lichen that grows on other plants; in lichens, used for lichens growing on bark or stems.

Filamentous — a type of fruticose lichen, with long and thin, hair-like branches (e.g. *Usnea*, *Bryoria*, *Alectoria*, etc.).

Foliose — made of flattened lobes, mostly spreading in two dimensions, which look different above and below, the upper surfaces having a layer of algae immediately beneath the cortex and lower surfaces often with attachment organs, e.g. rhizines. The lobe edges can be lifted from the surface they are growing on with a fingernail.

Fruiting Bodies or **Fruits** — informal but often-used terms in fungal biology to refer to sexually produced spore-bearing structures, including **apothecia**, **lirellae**, **perithecia** and mushrooms.

Fruticose — bushy or three-dimensional growth form with branches that look the same all the way round, due to having **photobionts** in a layer all the way around each branch.

Genus (plural: **genera**) — a principal taxonomic category that ranks above species and below family, and is denoted by a capitalised scientific name. Similar species are collected into genera if they are closely related.

Globular — rounded in three dimensions; usually roughly or irregularly spherical or globe-like.

Granular — made of tiny, rounded and smooth particles, individually visible at 10×; in soredia, used to contrast with fine or floury, which appear powdery.

Granule — rounded, strongly convex or sometimes flattened soredia-type structures for dispersal, with or without a cortex covering.

Habitat — the place or environment where an organism naturally or commonly lives and grows.

Heath — a low, tough vegetation type dominated by heathers and related plants, in acidic and low nutrient environments and on peaty soils, often with many *Cladonia* species between and below shrubs.

Holdfast — a strong single point attaching a lichen thallus to the surface it is growing on (its substrate); especially found in fruticose lichens like *Usnea* or *Ramalina*.

Hyphae — the threadlike, usually microscopic, cells of fungi.

Hypothallus — a mat of spongy fungal tissue on which lichenised parts grow; found especially in *Pectenia* species and visible at the tips of lobes as a thick weft of richly branching hyphae.

Isidia (singular: **isidium**) — projections of the upper thallus surface, continuous with the fungal cortex and containing algae, specialised for breaking off and dispersing. These may be finger-like, branched like tiny coral fragments, rounded or flattened.

Leprose — a thallus that is made up entirely of powdery grains (soredia).

Lichen fungi — fungi that form lichens.

Lichenised — a lifestyle or condition of certain fungi referring to the state with a symbiotic photobiont within its tissues, or the part of a lichen that contains both fungal and photobiont cells. Some lichen fungi grow in fungal-only zones at their growing edges; these are not lichenised.

Lime-rich — a substrate with calcium/magnesium carbonates in some proportion; used to describe soils or rocks, including limestone, marble, some sandstones, mortar, and cement.

Lirella (plural: **lirellae**) — a type of apothecium, elongate or linear in shape and often with black margins; characteristic of Script Lichens (see p. 84; Baker's Dozen Scripts and Commas).

Lobed Crustose — subtype of the **crustose** growth form, with lobe-like extensions that appear foliose towards the edge of the thallus but are still fully attached to the surface and cannot be lifted.

Lobes — the flattened, repeated parts, usually of a leafy lichen; measure these away from growing tips and branching points; see p. 39.

Lobules — small lobes, especially for vegetative fragmentation and dispersal, often arising as isidia-like structures, with a cortex.

Lowlands — the relatively warmer and drier southern and eastern parts of Britain and Ireland; compare **uplands**.

Margin — the rim of an apothecium (p. 34) or, sometimes, the growing edge of a thallus. An apothecium margin with internal algae looks like the thallus and is called a **thalline margin**; one with no algae usually differs from the thallus and is called a **non-thalline margin**.

Medulla — Loosely woven inner layer of a layered lichen below the photobiont layer, often cottony-white.

Moorland — open upland habitats dominated by heather and other dwarf shrubs, grasses and bog mosses on peaty, moist soils.

Mycobiont — the dominant fungal partner in a lichen, which usually forms the bulk of the biomass and the structure of the lichen; the scientific name of a lichen refers in a strict sense only to the mycobiont.

Non-thalline — referring to the margin of an apothecium without thallus tissue, so without a photobiont layer within (p. 34); also called **lecideine**. Often the same colour as the disc.

Nutrient-enriched — used when referring to lichen habitats and microhabitats that are enriched in nitrogen compounds from the addition of agricultural or road dust, fertiliser, bird droppings or sea spray.

Oceanic — temperate, humid climates with relatively small seasonal fluctuations in Britain and Ireland, near the Atlantic coasts.

Pendant — thalli are more than two times longer than wide; see *Usnea dasopoga*, p. 101.

Perithecia (singular: **perithecium**) — convex, closed, sexual spore-bearing structures (fruiting bodies) found in some lichenised fungi (e.g. the *Pyrenula* and *Verrucaria* groups and others), often coal black, sometimes sunken in the thallus; these are tiny and best interpreted at 10×, when you should be able to see the opening at the top for spore discharge; see *Pyrenula*, p. 227.

Photobiont — the photosynthetic partner of a lichen, fixing atmospheric carbon as a food source for the lichen; these can be either green algae or cyanobacteria – and sometimes both, in which case the cyanobacterium functions mostly as a nitrogen fixer.

Pin lichen — crustose lichen with stalked apothecia that look like small pins, p. 197.

Podetia (singular: **podetium**) — the formal name for the stalk of *Cladonia*.

Prothallus — the fungal tissue that precedes the development of a lichenised portion of a crustose lichen, sometimes visible as a white or black fringe of radiating fungal threads at the edge of a lichen, best seen when growing unimpeded by other lichens, or sometimes between the areoles.

Pruina — a fine, powdery crystalline coating on the upper surfaces of some lichens, especially on the cortex or apothecia, visible at 10× as a dusting like icing sugar; see *Physconia*, p. 177, or *Glaucomaria*, p. 215.

Pseudocyphellae — breaks in the upper or lower cortex for airflow, usually visible as raised pale patterns like white dots, lines or networks.

Pycnidia — asexual spore-producing structures, often visible as tiny regular black dots (10×) with pores at the tip, which can be difficult to see even at 10×.

Rhizines — hair or root-like attachment structures on the lower surface of leafy lichens. These may be simple, branched, fluffy or bottlebrush-like.

Silica-rich — rocks with relatively high silica content especially granite, gneiss, quartzite, slate and many sandstones. These are sometimes called siliceous or acidic rocks and, from the point of view of lichens, contrast with lime-rich or calcareous rocks.

Soralia (singular: **soralium**) — distinct clusters or patches of soredia deriving from breaks in the thallus surface, usually forming characteristic shapes depending on the species; may be round to elongate in outline and mounded, flat, or upturned (on lobe tips) in profile.

Soredia (singular: **soredium**) — powdery grains including both fungi and photobionts specialised for dispersal and arising from the lichen medulla; these have no cortex and therefore appear powdery *en masse* or matt-textured individually. Found either scattered across the surface of the lichen, or in localised, clearly distinct patches (**soralia**).

Sorediate — having **soredia**.

Squamules — small, foliose but minute scale-shaped structures, differentiated top from bottom, and arising and attaching to the substrate individually.

Squamulose — having **squamules**; a subtype of the foliose growth form.

Substrate — the surface on which a lichen directly grows.

Symbiont — a member of a stable association of two or more different organisms; the fungal partners and photobionts in lichens are both symbionts, as are cyanobacterial partners in nitrogen-fixing species.

Thalline — referring to the thallus, and therefore including photobiont cells; usually in the sense of 'thalline margin' of apothecia, in which the margin appears the same as the thallus.

Thallus (plural: **thalli**) — the vegetative body of a lichen, including both fungal and photobiont (algal or cyanobacterial) cells or tissues.

Tomentum — a short and usually even covering of fine hairs, like felt or velvet.

Uplands — the parts of the country that are cooler and wetter due to being further north or of higher elevation. Uplands are restricted to the highest elevations in the southern parts of England, but descend to sea level in northwest Scotland.

Zoned margin — the growing edge of some crustose lichens, forming concentric bands of different colours; see *Ochrolechia parella*, p. 205.

PHOTO CREDITS

Numbers refer to page numbers, and letters refer to positions on the page (Top, Middle, Bottom, Left, Right) or to panels in the Baker's Dozen and Rusty Reds sections. All photographs by Rebecca Yahr with the exception of the following:

Alan Watson Featherstone 12BL
Andy Acton 77C, 119TL, 137ML, 151TR, 151TL, 154TL, 233F
April Windle 75C, 75D, 90BL, 126 Col. 1: image 3, 138TL, 138TR, 176BR, 210BR
Barrie Hamill 65A, 123TR, 233C
Cairngorm House 9
Caz Walker 53 Row 5R, 75F, 77F, 207TL, 211TR, 224BR
Chris Cant 21BL, 21BR, 214TR
Claire Halpin 51 Row 1M, 59 Row 4L, 73I, 156MR, 209TR, 213ML
Dave Green 12BR
David Genney 11T, 18TM, 18TR, 18ML, 18M, 18MR, 18BL, 18BM, 21T, 36BR, 43TL, 43MR, 57 Row 3R, 57 Row 4L, 59 Row 1L, 75J, 77D, 88 Row 4: image 3, 89 Row 3: image 2, 95TL, 95TR, 101BL, 102MR, 109B, 112MR, 115TL, 116TR, 117TR, 118B, 119ML, 124TL, 124MR, 126 Col. 1: image 4, 126 Col. 2: image 4, 126 Col. 4: image 4, 127 Row 3: image 2, 127 Row 3: image 2, 134TL, 134MR, 135TR, 135ML, 136TL, 136TR, 137B, 138B, 142MR, 147TR, 154TR, 156TR, 156B, 174MR, 181TR, 202TL
David Purvis 33MR, 35B, 68BL, 171BR, 187 Row 5: image 3, Back flap Row 5, Back flap Row 8: image 4
Derek Christie 78B
Frances Stoakley 34ML, 35M, 36MR, 58B, 63 Row 2R, 68BR, 88 Row 3: image 3, 90BR, 118MR, 141TL, 159B, 172MR, 182MR, 183ML, 235TR
Graham Pyatt Front flap Row 2: image 4, 30BR, 33BR, 49 Row 4: image 2, 51 Row 4L, 61 Row 2L, 71K, 71M, 71N, 80B, 128 Row 4: image 1, 162B, 170BR, 173BL, 180TL, 180BL, 183TR, 187 Row 2: image 3, 187 Row 3: image 1, 187 Row 4: image 3, 189BR, 204B, 208BL, 210MR, 217ML, Back flap Row 7: image 4
Kristine Bogomazova 5, 19BL, 90TR, 149TR, 187 Row 5: image 4, 205TR, 228MR
Lynsey Wilson 14B, 25, 27TR, 36TR, 49 Row 3: image 3, 51 Row 3L, 59 Row 3R, 74A, 127 Row 1: image 3, 127 Row 4: image 2, 129TL, 163ML, 173TR, 187 Row 3: image 3, 205ML, 212TL, 212MR, 223TR
Marine Biological Association 88 Row 1: image 1, 96TL, 96MR
Mark Powell 16, 177BR, 193B, Back flap Row 2
Maxine Putnam 100BR, 225TL, 225ML, 225TR
Mike Simms 123BL, 123BR, 199ML
Mike Sutcliffe 61 Row 1: images 2 and 3, 89 Row 3: image 3, 94B, 95B, 120B, 132B, 140TL, 140MR, 143MR, 144B, 186 Row 3: image 4, 187 Row 1: image 2, 199TR, 233A
Nathan Chrismas 14T, 89 Row 2: image 2, 96TR, 97B, 116TL
Neil Bell 234BL
Neil Sanderson 52T, 128TR, 145BL, 175B, 224MR
Paul Cannon 53 Row 1M, 53 Row 4R, 57 Row 1R, 69F, 91R, 92MR, 99ML, 99B, 100TL, 100TR, 102TL, 105TR, 109TR, 111TR, 152TL, 154TR, 155TL, 155BR, 158MR, 180TR, 181ML, 187 Row 5: image 2, 223BL, 226MR, 233E
Pete Martin 61 Row 4M, 138MR, 149BR, 231TR, 231ML
Peter Upton 88 Row 1: image 3, 90TL, 90MR
Ray Woods 152MR
Rob Yaxley 164B
Sam Ebdon 10TL
Sandy Coppins 69I, 137TR, 144TL

Photo credits 247

ACKNOWLEDGEMENTS

Gaining the confidence to take on a project like this involves quite a journey, and we are very grateful for the patient teachers and friends who have shared their knowledge – sometimes about the same lichen, again and again! – and for the students who have asked probing and thought-provoking questions. This project has had a long incubation and many individuals have contributed in ways large and small, through conversations, field trips, queries and photographs (credited separately).

In particular, we thank Andy Acton, Judith Allinson, Peder Aspen, Kristine Bogomazova, Graham Boswell, Paul Cannon, Chris Cant, Nathan Chrismas, Derek Christie, Brian Coppins, Sandy Coppins, Andy Cross, John Douglass, Chris Ellis, Dave Genney, Olivia Gray, Anna Griffith, Claire Halpin, Barrie Hamill, Joe Hope, Amanda Jones, Les Knight, Pete Martin, Fay Newbery, Heleen Plaisier, Mark Powell, Steve Price, David Purvis, Maxine Putnam, Graham Pyatt, Neil Sanderson, Janet Simkin, Mike Simms, John Skinner, Cayla Smith, Dylan Stephens, Mark Stephens, Peter Upton, Caz Walker, Lynsey Wilson, April Windle, Ray Woods and Rob Yaxley. Special thanks go to Paul Cannon, David Genney, and Graham Pyatt who shared their curated photographic collections with us. The Marine Biological Association, Cairngorm House, Sam Ebdon, Dave Green and Alan Watson Featherstone have provided difficult-to-capture and stunning images of animals who interact with lichens. Mike Sutcliffe's immensely valuable website for learning lichens, managed by the British Lichen Society, has helped and continues to support lichen learners at all stages. We are grateful for the use of a selection of his images. Frank Dobson formed the foundation of learning lichens in Britain for us both, and we acknowledge the huge value of his books, fold-out guides and keys. We extend our heartfelt appreciation to Liz Campbell, illustrator-extraordinaire, for the careful rendering of lichen anatomy and for her accommodating and ever-helpful approach to this project.

We are grateful to all the members of the British Lichen Society who share knowledge freely and contribute records to the BLS, and we extend particular thanks to Brian Coppins, Janet Simkin and Mark Seaward who have built, maintained and managed the BLS Database. Janet Simkin has provided access to up-to-date BLS mapping data. Becky would also like to thank her early teachers Sylvia and Steve Sharnoff, Andrea Gargas, Bill Buck, Dick Harris and Ernie Brodo for providing a structure to lichen learning. We are very grateful to Alex Davey, Chris Ellis, Fay Newbery, Heleen Plaisier, Maxine Putnam, John Skinner, and Mark Stephens for reviewing drafts of the book and providing helpful and critical feedback and suggestions. Although we acknowledge the investment in us and the resources, conversations and relationships we've built from all of these people, we also accept any errors or inaccuracies as our own.

Special thanks go to the team at Bloomsbury, including Katy Roper, David Price-Goodfellow, Namrita Price-Goodfellow, Shane O'Dwyer and David Hawkins, for helping to turn ideas into the book we have wished for years that we had when teaching introductory classes.

Maps are drawn based on modelled British Lichen Society data from Colin Harrower, Oliver Pescott and Gary Powney at the Centre for Ecology and Hydrology, and they are warmly thanked for their work and collaboration.

We would also like to acknowledge our families, Tom Deacon and Chris, Simon and Wynn Ellis for being patient and relentlessly supportive with us while we worked on this project. Thank you for cups of tea, warming suppers, holding down the fort and taking up the slack. It is with you that we love to go out to explore the countryside; you make our adventures and expeditions second to none.

INDEX

abietina, Lecanactis 15, 197, 226
Abraded Camouflage Lichen 159
Acarospora 59, 229, 233
aculeata, Cetraria 92
adglutinata, Hyperphyscia 132
adscendens, Physcia 61, 72, 73, 180
aethalea, Buellia 6, 212
afrorevoluta, Hypotrachyna 71, 169
aipolia, Physcia 179
albescens, Lepra 79, 216, 217
Alectoria 87, 97, 101
alpina, Lichenomphalia 228
Alpine Bloodspot 5, 209
amara, Lepra 33, 49, 79, 216, 217, 221
amara, Pertusaria 217
ambigua, Parmeliopsis 51, 73, 175
Amorphophallus 13
amplissima, Ricasolia 137
Anaptychia 63, 73, 131, 176
androgyna, Ochrolechia 51, 53, 216, 217, 220
Antler Lichen 131
arbuscula, Cladonia 57, 67, 114
Arctic Saucer Lichen 220
Arctoparmelia 175
argena, Phlyctis 122
Arthonia 84, 223, 224
articulata, Usnea 53, 100
Aspicilia 203
Asterisk Lichen 223
Athallia 81, 194, 195
atlantica, Pectenia 53, 153, 154
atra, Arthonia 224
atra, Tephromela 35, 63, 204, 207, 208
atrata, Tremolecia 233
aurantia, Variospora 189
aurantiacum, Erythricium 237
aurella, Candelariella 193
aureola, Xanthoria 63, 139
auriforme, Collema 143
auriforme, Lathagrium 61, 78, 143

baeomyces, Dibaeis 214, 222
Baeomyces 57, 59, 87, 222

Bark Barnacles 213
Bay Moon Lichen 150
Beard Lichen 68–69, 91, 98–101
  Boreal 98
  Fishbone 101
  Inflated 99
  Red 91
  Warty 100
bellidiflora, Cladonia 107
bicolor, Bryoria 97
Big Browns 76
Bitter Wart 217
Black Eye Lichen 204
Black-eyed Susan 119
Black Seaweed Lichen 96
Blackthread Lichen 231
Blackwart Lichen 230
Blastenia 63, 81, 194, 195
Blennothallia 61, 144
Blistered Loop Lichen 169
Bloodspot, Alpine 5, 209
Bloody-heart Lichen 211
Blue Felt Lichen 153
Blue-grey Rosette Lichen 178
Bootstrap Lichen 95
Boreal Beard Lichen 98
britannica, Peltigera 138
Brown Beret Lichen 222
Brown Cobblestone Lichen 229
Brown-eyed Shingle Lichen 181
Browned Pixie Cup 94
Bryoria 27, 50, 87, 97
Buellia 6, 212
Bunodophoron 118, 119
Button Shield 168

caesia, Physcia 61, 73, 178
calcarea, Aspicilia 203
calcarea, Circinaria 8, 61, 203
Calcareous Rimmed Lichen 203
calcicola, Xanthoria 63, 139
calicaris, Ramalina 123
Calicium 197

Calogaya 80, 189
Caloplaca 80, 80-81, 188-190, 195, 196, 201
Camouflage Lichen 71, 133, 159
   Abraded 159
   Polished 133
   Shiny 160
   Warty 161
campestris, Lecanora 198, 207
canariensis, Sticta 21, 152
Canary Moon Lichen 152
candelaria, Polycauliona 141
candelaria, Xanthoria 141
Candelaria 55, 141, 191
Candelariella 15, 81, 191, 193, 196, 236
candelaris, Chrysothrix 15, 197, 226
candicans, Solenopsora 199, 200, 201
Candy Lichen 214
canescens, Diploicia 61, 199, 200, 201
caperata, Flavoparmelia 49, 167
carpinea, Glaucomaria 207
Cartilage Lichens 65
ceratina, Usnea 100
cervicornis, Cladonia 94, 102
Cetraria 57, 87, 92, 93
Cetrelia 166, 171
Chewing Gum Lichen 198
chlarotera, Lecanora 206
chlorophaea, Cladonia 57, 67, 102
chlorophylla, Tuckermannopsis 172
Chocolate Chip Lichen 134
christiansenii, Illosporiopsis 236
chrysophthalmos, Teloschistes 90
Chrysothrix 15, 197, 226
Cigarette Ash Lichen 215
ciliaris, Anaptychia 73, 131, 176
ciliata, Cladonia 113
cincta, Orchesella 10
cinnabarinum, Coniocarpon 223
Circinaria 8, 61, 198, 203
citrina, Caloplaca 196
citrina, Flavoplaca 55, 61, 81, 196
citrina, Pseudocyphellaria 151
Cladonia, Many-forked 110
T... 'one 182

Cladonia 14, 27, 29, 33, 48, 50, 56, 57, 66-67, 87, 92, 94, 102-114, 182, 184, 211, 235
Coastal Firedot 190
coccifera, Cladonia 27, 57, 67, 104
Collema 143-144
collina, Peltigera 53, 149, 158
Comma Lichen 84, 213, 223, 224, 227
Common Clam Lichen 183
Common Dust Lichen 218
Common Goldspeck 193, 196
Common Greenshield 167
Common Moon Lichen 150
Common Powderhorn 109
Common Script Lichen 224
Common Stippleback 157
Common Sunburst Lichen 139
Common Witch's Hair 101
concolor, Candelaria 55, 141, 191
confinis, Lichina 96
Coniocarpon 223
coniocraea, Cladonia 48, 67, 109
conizaeoides, Lecanora 16
conoplea, Pannaria 181
conspersa, Xanthoparmelia 59, 71, 163, 164
contorta, Circinaria 198, 203
Coral-crust, White 219
Coral Lichen 118
corallina, Lepra 59, 219
corallina, Pertusaria 219
corallinus, Marchandiomyces 236
Cornicularia 95
cornuta, Cladonia 109
cornuta, Usnea 53, 69, 99
Crab's Eye 205
crenularia, Blastenia 63, 81, 194, 195
Cresponea 197, 210
crinitum, Parmotrema 171
Crinkled Snow Lichen 124
crispa, Blennothallia 61, 144
crocea, Solorina 57, 134
Crottle, Smoky 162
Cryptic Firedot 201
cuspidata, Ramalina 65, 120, 121
cyanoloma, Pectenia 153, 154
cyathoides, Fuscidea 11, 83

cylindrica, Umbilicaria 59, 156

dactylophyllum, Stereocaulon 117
dasopoga, Usnea 50, 69, 101
decipiens, Calogaya 80, 189
dendritica, Phaeographis 224, 225
Dermatocarpon 61, 157
Desperate Dan 171
Diarthonis 223
Dibaeis 87, 214, 222
digitata, Cladonia 105
Diploicia 61, 199, 200, 201
Diploschistes 12
dispersa, Myriolecis 55, 82, 207
distorta, Physconia 177, 179
Dotted Ribbon 23
Dragon Horn 108
dubia, Physcia 178
Dust Lichen, Common 218

Eared Jelly 143
elaeochroma, Lecidella 83, 210
elegans, Graphis 224
elegans, Rusavskia 140, 189
Elf Ears 184
Enchylium 10, 144
ericetorum, Icmadophila 214
ericetorum, Lichenomphalia 19, 228
Erythricium 237
esperantiana, Usnea 99
Evernia 15, 49, 50, 87, 123, 130, 131
evolutum, Stereocaulon 117
exasperata, Melanohalea 159
exasperatula, Melanohalea 32

False Reindeer Moss 112
Fanfare of Trumpets 23
farinacea, Ramalina 33, 48, 65, 122, 123, 130
fastigiata, Ramalina 23, 65, 123
Felipes 84
Felt Lichen 77, 154
    Blue 153
fimbriata, Cladonia 102
Firedot 80–81, 103, 194–195, 232
    Coastal 190

Cryptic 201
    Limestone 188
    Mealy 196
Fishbone Beard Lichen 101
flavescens, Caloplaca 188
flavescens, Variospora 30, 55, 61, 80, 188
flavicans, Teloschistes 90
Flavocetraria 57, 87, 124
flavocruenta, Porpidia 233
Flavoparmelia 49, 55, 70, 167, 235
Flavoplaca 55, 61, 63, 81, 190, 196, 232
Flecked Pox 227
floerkeana, Cladonia 67, 106
florida, Usnea 69
Foam Lichen, Variegated 116
fragilis, Sphaerophorus 118
fraxinea, Ramalina 65, 123
friesii, Xylopsora 183
frigida, Ochrolechia 57, 220
Frilly Lettuce Lichen 172
Fringe Lichen 39, 72–73, 176
Fringed Rock Tripe 156
Fringed Rosette Lichen 180
Frost Lichen 39, 72–73
    Grey 177
Frullania sp. 234
fuciformis, Rocella 120
fuliginosa, Melanelixia 71, 159, 160
fuliginosa, Sticta 150
furcata, Cladonia 92, 94, 110
furfuracea, Pseudevernia 51, 130, 131
fuscata, Acarospora 59, 229
fuscescens, Bryoria 27, 50, 97
Fuscidea 11, 83, 210, 237

gangaleoides, Lecanora 204, 207
gelatinosum, Leptogium 142
gelatinosum, Scytinium 142
geographicum, Rhizocarpon 11, 58, 59, 95, 192
glabratula, Melanelixia 133, 159
glauca, Platismatia 51, 162, 172, 236, 238
Glaucomaria 207, 215, 216, 219
globosus, Sphaerophorus 50, 59, 90, 118, 119, 211
Gold Dust 197

Golden Eye Lichen 90
Golden Hair Lichen 90
Golden Specklebelly 151
Goldspeck, Common 193, 196
    Lobed 191
gracilis, Cladonia 94
Granite Speck 208
Graphis 53, 224, 225, 227
Green Reindeer Lichen 114
Green Satin Lichen 137
Greenshield, Common 167
Grey Frost Lichen 177
Grey Reindeer Lichen 114
grisea, Physconia 40, 49, 55, 177, 178
Gritty British Soldiers 106

Haematomma 209
Hammered Shield 165
Haugania 233
Heath Navel 228
Heath Thorn 111
Hoary Rosette Lichen 179
holocarpa, Athallia 81, 194, 195
Hooded Tube Lichen 173
horizontalis, Peltigera 147
Horsehair Lichen 97
hudsoniana, Lichenomphalia 228
humilis, Cladonia 66, 67
Hydropunctaria 63, 190, 230, 232
hymenea, Pertusaria 207, 213, 217, 221
hymenina, Peltigera 28, 74, 145
Hyperphyscia 132
Hypocenomyce 51, 183
Hypogymnia 49, 51, 71, 165, 173, 174
Hypotrachyna 51, 53, 71, 165, 169–170, 235

Iceland Moss 93
Icmadophila 214
Illosporiopsis 236
incana, Lepraria 55, 218
incurva, Arctoparmelia 73, 175
Inflated Beard Lichen 99
intricata, Lecanora 208
islandica, Cetraria 57, 87, 92, 93

jeckeri, Punctelia 55, 166
Jelly Lichen 10, 37, 78, 142–44
    Eared 143
    Soil 144
Jellyskin, Rose-petalled 142

Kidney Lichen 77
    Mustard 158
Kuettlingeria 200–201

laciniatula, Melanohalea 71, 159
Laetisaria 237
laevigata, Hypotrachyna 51, 53, 170
laevigatum, Nephroma 158
lambii, Placopsis 202
Lasallia 135
Lathagrium 61, 78, 143
Lecanactis 15, 197, 226
Lecanora 10, 16, 34, 82, 198, 204, 206–208, 210, 213, 236
lecanorinum, Rhizocarpon 192
Lecidea 59, 83, 212, 233
Lecidella 34, 83, 210
Lecidella Lichen 210
lepadinum, Thelotrema 53, 213, 227
Lepra 33, 49, 59, 79, 216, 217, 219, 221
Lepraria 55, 218
leptalea, Physcia 179, 180
Leptogium 142
leucopellaeus, Felipes 84
leucophlebia, Peltigera 75, 138
lichenicola, Laetisaria 237
lichenoides, Scytinium 142
Lichenomphalia 19, 228
Lichina 63, 78, 96
lightfootii, Fuscidea 210
limbata, Sticta 149
Limestone Firedot 188
Lipstick Pixie Cup 104
Litmus 120
Little Brown Map Lichen 212
Lob Scrob 148
Lobaria 10, 21, 53, 76, 77, 136, 137
Lobarina 53, 77, 148
Lobed Goldspeck 191

Loop Lichen 71, 169, 170
    Blistered 169
    Smooth 170
loxodes, Xanthoparmelia 63, 161
lucida, Psilolechia 196
Lung Lichen 77
Lungwort 136
    Textured 148
lygaea, Fuscidea 237

macrospora, Pyrenula 227
mamillata, Anaptychia 176
Many-fingered Powderhorn 105
Many-forked Cladonia 110
Map Lichen 11, 212
    Little Brown 212
    Yellow 58, 95, 155, 192
Marchandiomyces 236
marina, Flavoplaca 63, 190, 232
maura, Hydropunctaria 63, 190, 230, 232
maura, Verrucaria 232
Mealy Firedot 196
Mealy Pixie Cup 102
Mealy Shadow Lichen 132
medians, Candelariella 191
Melanelia 71
Melanelixia 49, 55, 71, 133, 159, 160, 238
melanocarpum, Bunodophoron 118, 119
Melanohalea 32, 71, 159
melinodes, Porpidia 233
membranacea, Peltigera 74, 75, 146
Membranous Pelt Lichen 146
Menegazzia 71, 174
Metzgeria sp. 234
miniatum, Dermatocarpon 61, 157
Moon Lichen 76, 150
    Bay 150
    Canary 152
    Common 150
    Powdered 149
Moss, False Reindeer 112
    Iceland 93
    Oak 130
mougeotii, Xanthoparmelia 71, 164, 175
mucosa, Wahlenbergiella 62, 63, 232

muralis, Lecanora 198
muralis, Protoparmeliopsis 55, 198
Mustard Kidney Lichen 158
Mycoblastus 53, 83, 211
Myriolecis 55, 82, 207

Navel, Heath 228
Nephroma 77, 158
Netted Rock Tripe 156
nigrescens, Verrucaria 61, 199, 203, 230, 231
nigricans, Alectoria 97
nigrum, Placynthium 61, 230, 231
nivalis, Flavocetraria 57, 124
Normandina 53, 184
normoerica, Cornicularia 95

Oak Moss 130
Ochrolechia 51, 53, 57, 59, 63, 79, 205, 216, 217, 220
ochroleucum, Haematomma 209
oederi, Haugania 233
Old-wood Lichen 226
olivetorum, Cetrelia 166, 171
omphalodes, Parmelia 59, 70, 160, 162, 163
Ophioparma 59, 209, 236
orbicularis, Phaeophyscia 55, 132, 177
Orchesella 10
Orchil 120

Pannaria 53, 154, 181
parella, Ochrolechia 59, 63, 79, 205
parietina, Xanthoria 10, 17, 40, 49, 55, 63, 72, 139, 140, 180, 235
parile, Nephroma 158
Parmelia 7, 15, 19, 28, 33, 49, 51, 55, 59, 70, 73, 160, 162–163, 165, 211, 235, 236, 237
Parmelina 165, 168
Parmeliopsis 51, 175
Parmotrema 49, 71, 165, 170, 171, 172, 238
pastillifera, Parmelina 168
Pectenia 53, 77, 153–154, 181
Pelt Lichen 74–75, 146, 148
    Membranous 146
    Ruffled Freckle 138
    Smooth 145

Peltigera 20, 21, 28, 53, 74–75, 75, 138, 145, 146–147, 148, 149, 158
Pepper Pot 221
Peppered Rock Shield 164
perlatum, Parmotrema 49, 171
pertusa, Lepra 49
pertusa, Pertusaria 217, 221
Pertusaria 79, 207, 213, 217, 219, 221
Petalled Rock Tripe 155
Phaeographis 224, 225
Phaeophyscia 55, 73, 132, 177
Phlyctis 122
phycopsis, Rocella 120
Physcia 28, 49, 55, 61, 72, 72–73, 139, 140, 178–180, 236, 237
Physconia 39, 40, 49, 55, 73, 140, 177, 178, 179
physodes, Hypogymnia 49, 51, 173, 174
Pincushion Sunburst Lichen 140
Pink Bull's Eye 202
Pixie Cup 66–67
    Browned 94
    Lipstick 104
    Mealy 102
    Rosette 103
Placopsis 202
Placynthium 61, 230, 231
Platismatia 51, 71, 162, 172, 236, 238
Pleated Lichen, White 200
plumbea, Pectenia 154
pocillum, Cladonia 67, 102, 103
Polished Camouflage Lichen 133
polycarpa, Polycauliona 140
polycarpa, Xanthoria 140
Polycauliona 63, 140, 141, 190–191
polydactyla, Cladonia 50, 104–105, 211
polyphylla, Umbilicaria 59, 155
polyrrhiza, Umbilicaria 155
polytropa, Lecanora 82, 208, 236
Poppadom Lichen 135
Porpidia 59, 83, 215, 233, 237
Port-hole Lichen 174
portentosa, Cladonia 33, 57, 67, 113
Powdered Moon Lichen 149
Powdered Ruffle Lichen 171

Powderhorn Lichen 66
    Common 109
    Many-fingered 105
Powder-tipped Rosette 178
Powdery Saucer Lichen 216
Pox Lichen 227
    Flecked 227
praetextata, Peltigera 53, 147
praetextata membranacea, Peltigera 75
premnea, Cresponea 197, 210
proboscidea, Umbilicaria 156
Protoblastenia 61, 195
Protoparmeliopsis 55, 198
prunastri, Evernia 49, 50, 123, 130, 131
Pseudephebe 92, 95
Pseudevernia 51, 87, 130, 131
pseudocorallina, Pertusaria 219
Pseudocyphellaria 77, 151
Psilolechia 196
pubescens, Pseudephebe 92, 95
pulchella, Normandina 53, 184
pulmonaria, Lobaria 10, 53, 76, 77, 136
Punctelia 39, 49, 55, 70, 165, 166
pusilla, Calogaya 189
pustulata, Lasallia 135
pygmaea, Lichina 63, 96
Pyrenula 227
pyxidata, Cladonia 102, 103

radiata, Arthonia 223
Ramalina 23, 33, 48, 63, 64, 65, 87, 120, 121–123, 130
rangiferina, Cladonia 57, 114
rangiformis, Cladonia 110, 112
Red Beard Lichen 91
reductum, Rhizocarpon 83, 212
Reindeer Lichen 66–67, 113, 124
    False 112
    Green 114
    Grey 114
reticulatum, Parmotrema 171, 238
revoluta, Hypotrachyna 169
Rhizocarpon 11, 30, 58, 59, 83, 95, 192, 212, 233
Ricasolia 53, 77, 137

Rim Lichen 82, 206, 221
Ring Lichen, Sorediate 175
Rocella 120
Rock Lichen 11
Rock Tripe Lichen 95, 135
    Fringed 156
    Netted 156
    Petalled 155
Rose-petalled Jellyskin 142
Rosette Lichen 39, 72–73
    Blue-grey 178
    Fringed 180
    Hoary 179
    Powder-tipped 178
Rosette Pixie Cup 103
Rotten Oranges 195
rubicunda, *Usnea* 69, 91, 100
rubiginosa, *Pannaria* 53, 181
rufescens, *Peltigera* 75, 145
Ruffle Lichen 171, 172
    Powdered 171
Ruffled Freckle Pelt 138
rufus, *Baeomyces* 57, 59, 222
runcinata, *Anaptychia* 63, 73, 176
rupestris, *Protoblastenia* 61, 195
rupicola, *Glaucomaria* 215, 216, 219
Rusavskia 140, 189

saccata, *Solorina* 134
Salted Shield 163
sanguinarius, *Mycoblastus* 53, 83, 211
sarmentosa, *Alectoria* 101
Saucer Lichen, Arctic 220
    Powdery 216
saxatilis, *Parmelia* 33, 51, 59, 162, 163, 165, 211
saxicola, *Caloplaca* 189
scalaris, *Hypocenomyce* 51, 183
Script Lichen 84, 213, 224, 225, 227
    Common 224
scripta, *Graphis* 53, 224, 227
scrobiculata, *Lobarina* 53, 77, 148
Scytinium 142
Sea Ivory 121
Seaweed Lichen, Black 96
Shaggy Strap Lichen 122

Shield Lichen 39, 52, 70–71
    Button 168
    Hammered 165
    Peppered Rock 164
    Rock 71
    Salted 163
    Speckled 166
Shiny Camouflage Lichen 160
Shrek's Ears 23, 123
Shrubby Sunburst Lichen 141
siliquosa, *Ramalina* 63, 65, 120, 121
sinopica, *Acarospora* 229, 233
sinuosa, *Hypotrachyna* 71, 170
Smoky Crottle 162
Smooth Horn Lichen 94
Smooth Loop Lichen 170
Smooth Pelt Lichen 145
Soil Jelly Lichen 144
Soldiers, Gritty British 106
    Toy 107
Solenopsora, Limestone 199
Solenopsora 199, 200, 201
Solorina 57, 134
soredians, *Flavoparmelia* 55, 167
Sorediate Ring Lichen 175
spadicea, *Diarthonis* 223
Specklebelly Lichen 77
    Golden 151
Speckled Shield 166
Sphaerophorus 50, 59, 87, 90, 118, 119, 211
Spiny Heath Lichen 92
squamosa, *Cladonia* 67, 108
Starry Scribbles 225
stellaris, *Physcia* 179
Stereocaulon 11, 59, 87, 116–117, 134
Sticta 21, 53, 76, 149–150, 152
Stippleback, Common 157
strepsilis, *Cladonia* 182
String of Sausage Lichen 100
subaurifera, *Melanelixia* 49, 55, 133, 159, 238
subcervicornis, *Cladonia* 29, 67, 182
subfloridana, *Usnea* 48, 50, 69, 98
subrudecta, *Punctelia* 49, 55, 166
sulcata, *Parmelia* 7, 49, 55, 70, 163, 165
sulphurea, *Lecanora* 204, 208

Sunburst Lichen 72
   Common 139
   Pincushion 140
   Shrubby 141
Supersized Crusts 79
sylvatica, Sticta 53, 150

Tar Lichen 232
tartarea, Ochrolechia 79, 205, 216
taylorensis, Hypotrachyna 170
teicholyta, Caloplaca 201
teicholyta, Kuettlingeria 200, 201
Teloschistes 90
tenax, Collema 144
tenax, Enchylium 144
tenella, Physcia 49, 55, 72, 180
Tephromela 35, 63, 204, 207, 208
terebrata, Menegazzia 174
Textured Lungwort 148
thallincola, Caloplaca 190
thallincola, Variospora 63, 190, 232
Thamnolia 57, 87, 115, 134
Thelotrema 53, 213, 227
tiliacea, Parmelina 168
torrefacta, Umbilicaria 155
Toy Soldiers 107
Tremolecia 233
Trentepohlia 223, 235
tuberculosa, Porpidia 59, 215, 237
tubulosa, Hypogymnia 51, 173
Tuckermannopsis 172
Two-tone Cladonia 182

Umbilicaria 59, 95, 135, 155–156, 157
uncialis, Cladonia 57, 67, 111

Usnea 15, 26, 48, 50, 53, 68, 68–69, 87, 91, 98–101, 208, 238

Variegated Foam Lichen 116
Variospora 55, 61, 63, 80, 188–190, 232
ventosa, Ophioparma 59, 209, 236
vermicularis, Thamnolia 57, 115, 134
Verrucaria 61, 199, 203, 230, 231, 232
verruculifera, Polycauliona 63, 190, 191
verruculifera, Xanthoparmelia 71, 161, 164
vesuvianum, Stereocaulon 11, 59, 116, 117
virens, Lobaria 137
virens, Ricasolia 53, 77, 137
viride, Calicium 197
vitellina, Candelariella 15, 193, 196, 236

Wahlenbergiella 62, 63, 232
Warty Beard Lichen 100
Warty Camouflage Lichen 161
wasmuthii, Usnea 26, 69, 98
White Coral-crust 219
White Pleated Lichen 200
Whitewash Lichen 122
Whiteworm Lichen 115
Wine-gum Lichen 83, 212
Witch's Hair, Common 101

Xanthoparmelia 59, 63, 71, 161, 163, 164, 175, 235
Xanthoria 10, 17, 40, 49, 55, 63, 72, 139–141, 180, 235
Xylopsora 183

Yellow Map Lichen 58, 95, 155, 192